TWELVE
SHEEP

First published in Great Britain in hardback in 2024 by
Allen & Unwin, an imprint of Atlantic Books Ltd.

Photograph on p. 81 © John Connell
Image on p. 143: *Anguish* (Angoisses) (1879), August Friedrich
Albrecht Schenck, engraving, National Gallery of Victoria,
Melbourne. Purchased, 1880. Reproduced with kind permission of
the National Gallery of Victoria, Melbourne
Image on p. 147: *Shepherd and Sheep* © Erich Hartmann/Magnum Photos.

A CIP catalogue record for this book is available from the British Library.

Hardback ISBN 978 1 80546 190 6
E-book ISBN 978 1 80546 191 3

Allen & Unwin
An imprint of Atlantic Books Ltd.
Ormond House
26–27 Boswell Street
London WC1N 3JZ

www.atlantic-books.co.uk

Printed and bound by CPI (UK) Ltd, Croydon CR0 4YY

10 9 8 7 6 5 4 3 2

MIX
Paper | Supporting
responsible forestry
FSC
www.fsc.org
FSC® C171272

TWELVE
SHEEP

Life lessons from a lambing season

JOHN CONNELL

ALLEN&UNWIN

Contents

1 We must take a stake in the future 1

2 Home is where the heart is 15

3 Walking is good for the soul 29

4 In waiting, we grow 49

5 We must keep wonder in our minds 59

6 Labour is unavoidable 75

7 We must love our home 89

8 Death is part of life 113

9 All mothers are a link to the great mother 125

10 We must keep beauty in our minds 137

11 Health is wealth 151

12 Love is what you need 165

Epilogue 173

For Bríd
Viv and Oliver
Thank you for the words

He tends his flock like a shepherd:
He gathers the lambs in his arms
and carries them close to his heart;
he gently leads those that have young.

Isaiah 40:11

I long for scenes where man hath never trod
A place where woman never smiled or wept
There to abide with my Creator, God,
And sleep as I in childhood sweetly slept,
Untroubling and untroubled where I lie
The grass below – above the vaulted sky.

John Clare – 'I Am!'

LESSON 1

We must take a stake in the future

We try and live simply but the world is complex. It has always been this way.

It is early autumn and I am standing in the sheep shed of our farm. Before me stand twelve sheep. They are, to be precise, twelve hoggets, the name we give to maiden females. These twelve ladies are mine. I have bought them from my parents with the money I earned from my words, from my books. I am a shepherd for the first time in my life. I am in the twilight of my youth and the budding of my middle age. I am older than Christ when he died and the same age as Buddha when he attained enlightenment. Both figures have walked beside me for so many years now. They have been part of my continuance in ways I think that count for some good, though I'm no sage.

I've known sheep for seven years now. Seven years as a farmhand, seven years as a midwife and seven years, at times, as an undertaker. It has been a long

3

apprenticeship. I came home to Ireland from Australia to try my hand at being a writer, but in the process, I became a farmer. It happened naturally: it began with the sheep. It has been a sojourn into the earth and its creatures, albeit one in which I have never been an owner, never before as a farmer in my own right.

I have bought the twelve animals for many reasons but perhaps *the* one, the most important, is that they are a stake in the future but sheep also challenge you to live in the now. I like this mission. I must be ready for both situations, and as I look at the girls in front of me, I come to think that this is the right thing for me to do. The right journey to undertake.

Sheep are earthly creatures. They eat, they live, they die: they are wholly of this world. The sheeping business is part of this journey of life. Our work with the animals provides food for the people of the cities and towns of this world; their wool, though not of much monetary value to us, provides the raw material for clothes, for warmth in these cold times.

The girls will live in our fields; they have run with our blue Texel ram and come to be in lamb. It is this journey that interests me. This journey from youth to motherhood. The coming of the lambs will drop life on me and maybe there will be a wisdom in that. A learning in that. I am a man in search of new life: it will be brought from the wombs of these creatures. In the lambs, I set my aims. They will be my goal, and their passage to this world my prize.

I have worked cattle and horses but there is something that is calling me in these sheep. In their quiet nature upon this earth. In their nibbling over grass, their gentle walks upon our soil. The sheep give me calm; they ease a busy mind with their approach to the world. The sheep wants nothing more than to be a sheep and maybe I can learn something in that simple wish.

In working with sheep as I do now, I know that there will be labour. Hard work is not something I will shy away from. There will be feeding and probing. Dipping and shearing. All of this is the way of a shepherd. It is physical, intensive work. Work that demands a strong body or a body that is willing to be put through its paces. To work with animals needs a factor of strength and perseverance. Of course there, too, will be life and death. And perhaps in some queer way, heaven and hell in the days that come.

Though the work is only now beginning, I can say that these sheep have already saved me. They have brought me back from an edge. Before the twelve, I was suffering from what I can only call a fatigue, one that was perhaps soulful as much as physical. For six months before the sheep, I was empty and worn out. I had finished a book and found myself spent. There was, I reasoned, no more to say; perhaps I had said it all already. What do writers do when the words will no longer come? I did not know. It was a new threshold, a dark threshold.

I battled a tiredness such as I had not felt in many long years, and though I slept, I could not rest. My wife called it burnout; I called it a soul tiredness. I tried to write book after book but nothing came and then the stories dwindled down to a trickle and the well, as I had known it to be before, so full and overflowing, was dry and empty. It made me sad for the first time in a long time. It was not a depression, a dog I had danced with before, but there was melancholy and it wore me out all the same. The sadness felt harder than depression, as it bred a loneliness in my body and my mind that I did not want or like. The sadness and tiredness sapped away my creativity and for a writer that was like losing a part of myself, a much needed and valued limb. I did not sleep well. I did not rest well.

I lived for a time in this vast chasm. My wife and I bought a house. I helped a friend through the deaths of his young twin daughters after a tragic accident. And I thought seriously about leaving the land, of returning to a city and becoming an urban person once again. I missed the challenge of the city: in the quietness of the countryside, in which I had heretofore found inspiration, now I found nothing. It was a hard time, my empty time.

After a period, I farmed and walked the earth and gradually slowed down my life and, in many ways, my soul too. I removed myself from unwanted social outings and began to meditate again on the nature of life. I thought of the bards of the land, from the English pastoral poet John Clare to the environmentalist Rachel Carson. I remembered Henry David Thoreau's great line from *Walden*: 'Live in each season as it passes, breathe the air, drink the drink, taste the fruit and resign yourself to the influence of the earth.' There was something in that notion of 'the influence of the earth': maybe with this soul weariness I had not been in contact with her. I wondered if my own land could be a personal Walden, if I could truly claim my natural inheritance again after it seemed to be lost. To find or discover a personal Walden is to find the source of solace in a weary world. Home can be that place, a land of prisms through which we can see better our true inheritance. When we discover our personal Walden, we refract not just light but life itself.

One day, while reading, I came across a line by John Clare which said that he found the poems he was so beloved for in the fields and wrote them down. That struck me deeply and truly. Maybe the fields were the answer I was looking for? Clare knew something of sadness and longing, of belonging and not belonging. He had been hailed once as the great peasant poet and then cast aside when fashions changed in London, and he had ended his days in an asylum.

Clare, Thoreau and Carson; they spoke to me in new ways, and though their words are now old, I found fresh insight in them. I wondered in that time, and in the wondering, questions came. Into the living sea of waking dreams, as Clare put it, I found newness. The bards of the land were all in their quiet unhurried ways teaching me. In this space, I asked myself what I was doing here. It is, I think, a question we must all work out ourselves at some point in our lives. From the company director to the builder's labourer, we all face the question of whether we should maintain or change, if we should evolve or dissolve into something else. It was a time of great thought, a time when sadness went hand in hand with tiredness. Maybe in melancholy we can find new ways of looking at the world? I do not truly know.

After a time in the slowness, something happened. An idea came to me; the notion of the twelve sheep descended upon me. It came, as all great personal revelations do: quietly, without pomp or ceremony. It was not some Shakespearean apparition – there were no ghosts hovering over me, no Banquo moment. Rather it was an internal vision, I suppose, one where I could see myself with twelve sheep, walking upon the earth as a shepherd in the cathedral of nature. It came gently, like the blossom of the yellow winter rose had come to our garden when I was a child. It was a quiet hegemonic thought that refused to go away. Ideas, I think now, do not give, or rather are not given, their full virtue. A new idea is a thing to be treasured; it doesn't matter if it is original or not. A personal idea is new to oneself. That is all that counts. The idea, my idea, was a simple thing: to buy the twelve sheep, to follow their lives and to learn from them; and in that idea, I was reborn. The fatigue, the soul tiredness, lifted to be replaced with a fresh sense of purpose. My existence had an aim and my incompleteness was ended. I let go of my sadness. It was a new-found freedom. As Henri Nouwen, the writer, said: 'Most of us have an address but cannot be found there.'

I break from my story and look at these sheep, my sheep, again. These animals have been stapled to my heart, to the bulletin board of life; they have made the route home for me once more. Already, they have taken me out of myself or, rather, brought me back to myself. In the work of the sheep, *I* have a new way to work. I am in a new landscape, the *tierra* of my reunification.

Now, I fill a bucket of nuts and pour it into the small feeder in the shed and the twelve sheep come forward to eat. I rub their heads and feel their shoulders. They are twelve strong girls, twelve true addresses. One can judge a sheep by the meat upon its carcass. We grade them by looking at their straight backs and udders and make a mental picture of their features. I learned this way of seeing from my father, from our neighbours and from my studies. It is a new way of looking at an animal. A learned way that I have perfected in my apprenticeship.

Twelve is, I think, a good number, a hearty number. There are twelve months in the year and twelve signs in the zodiac (one of which is the ram); indeed, there were twelve apostles to our Lord. Hercules, when he carried out his mythical labours, completed twelve tasks. In the majesty of numbers, they seemed the right digits.

I've known these hoggets since birth and they know this farm. They were born here. Before the girls, there were other sheep – ones at the livestock mart, others belonging to neighbours, even some from the internet. We carried out research trips, we saw sheep aplenty, heard stories from an auctioneer in Leitrim about shearing sheep for a famous celebrity and met traders and day men, but none fit the bill. Some were skinny, some with bad teeth and broken mouths (a sign of age in sheep), while others just didn't seem to hit the sweet spot for a buyer. There are 3.7 million sheep in Ireland spread out over 35,000 flocks. There are sheep for sale every day but none so storied as these twelve.

It was my father who suggested we keep them on after our research trips had come to nought. It was my mother who said they should stay. They are from the old ram, and in buying them from my parents, they have not been uprooted from our home nor removed from this patch of earth. If I had not bought them, they would have been broken up, dispersed to marts and sale yards. Some might not have been kept for breeding and ended up under a butcher's knife or on a slaughterhouse floor. I did not save them from slaughter, as all things end that way at some point, but by my actions I kept them on our farm, and for that I am glad. In their genes will live on new sheep, new adventures.

They are Suffolk crosses with good breeding, their black faces revealing their ancestry back to the start of our days on this farm. We began with Suffolks and learned the ways of sheep with this breed. I like the Suffolks: they date from the 1700s, when Norfolk Horn ewes were crossed with Southdown rams in England. The first flock came to Ireland in 1891. They have never left. They are all over this island now and rightly call it home. They are suited to our lowland farms. They can grow in this landscape. They are a proud and, to my eye, a noble breed, their strength and size matched by their maternal qualities.

The girls are my best hope of learning life's lessons, for in them are contained both my beginner's mind and my guiding hand. The lessons, I think, will not all come at once: they will be peppered through the season in the ups and downs, in the quiet and busy moments. Perhaps I hope to learn most of all how to be alive again after my period of inaction. These sheep are the route to my 'clay soul' as John O'Donohue, the Irish poet and philosopher, called it. One cannot raise sheep to be wise nor train them to be brave, but in their quiet way, they are teachers. They know where the good earth lies, where the best shelter is to be found. To me, they are like little Buddhas wrapt in white fleeces, calm and serene, dropping wisdom slowly, or perhaps more fitting to this landscape, like the monks upon Skellig Mhichíl in the south of the nation long ago.

The sheep jostle and push one another now for the last of their nuts and I watch them in their different personalities. To an outsider these sheep may all seem the same, but if we take the time to look and observe them, they are as different as we are from each other. There are bullies and battlers in the group. There are girls who will dominate the others to get the best position for eating. I will get to know their personalities over this season. I hope they will be good sheep. I bow to them now for they deserve my awe. I'm no rancher, with my twelve ladies, but they are tying me anew to this land of ours. They and their bellies are full of life that holds such promise.

Behind me now, I hear my father walking up the shed. He is happy to see me take on this mission. It was not that long ago that we fought over farming and life choices but those times are past: we see each other now as individuals. I am a writer and he respects that. He is a farmer and I respect his knowledge. He is wise and calm and, looking at the sheep, he produces his old, practised phrase, adapted to this moment: 'There'll be twelve stories to be told before the season is done.'

In a way, I think the first lesson has come already: to grab the future and let go of the past. This book, short as it might be, will be a record of this time in the life of these sheep and this land and living of ours. A menagerie of the trials and tribulations. A story of when the earth speaks, and we listen. I have taken this stake in my future. It is my first lesson, come what may. I am ready to learn.

LESSON 2

Home is where the heart is

ome is a song to which I am still learning the words. There is the music of the fields, the birds, the meadows and the yet unheard notes of all of creation itself. I think I will be singing these choruses all my life in so many different ways. It is my one true canto.

Home is important to me, not just for holding farm animals: it is the holder of a true world. A world somehow still precious. When I think of home, I think of its safety: that in the universe of the fields we can perhaps divine true meaning, as Hermann Hesse, the German-Swiss writer, said, in our 'blood whispers'. Hesse called himself a seeker in that quote but said he looked no longer in books or stars but to himself. When I look inward to my own voice, part of that voice is made of this place. There under the sky of home I can rest a true rest in this small universe. When I seek, I look to home first: it is where so many answers can be found.

The story of my home dates back to my great-grandfather, who came here from the nearby village of Ballinamuck, where the infamous 1798 rebellion ended in a field, where French and Irish men lay slain and defeated. The village recalls this battle still, and the visit of our French brothers, in its garden of remembrance, and the tragedy of the execution of the Irish rebels in the nearby hamlet of Ballinalee. Great-grandfather Connell married into a family called the Reillys. He was a farmer and a tenant to Lord Granard. He walked these fields long before I was even thought of. I have looked him up on the census and seen his title, his profession, and understood that we are connected by this same patch of earth.

My mother's people came from Bunlahy, a parish nearby. Her people date to the 1700s and came, I'm told, originally from Cavan, north of here, the land of the Breffni tribe. As to when they came south, we do not know and those that did are long gone.

The family has known these homelands all our lives but the farm we work on is a new thing, a creation of my father's and mother's dreams and will. Apart from the home paddocks, the other lands belonged to other families and other people, but slowly, over the course of thirty years, my parents have built the farm into our holding. We call it Birchview. From the Soran river to the hills of Clonfin right down to the Camlin river, it is a universe all of our own making. Others have left their handprints on this land, their names for fields and places remaining long after their departure, telling us something of these people of history. About how they lived and perhaps what they dreamed of. We are the heirs of those dreams. Some of the fields have known our hands and they have been reshaped and repurposed and given new titles, like The Garden (though it has only known vegetables once) and The Big Meadow (though it is usually eaten during summer and seldom cut for silage). Despite the labours and the new names we, too, know that we are but passing through. We are this generation's custodians; as to how long we will hold the mantle, that is up to the gods.

Land is important business here. Land, as we say, runs deep. Maybe it is because we had so little control over our land and destiny for hundreds of years of colonialism. I am not sure, but land is in our psyche. It is both a dream and a living thing. A place of two worlds.

I have been at home now for seven years, and daily I learn the lesson of what it means to be centred in a place. They say that every seven years we become new people, that our skin, our hair, our cells replace themselves wholly so that we emerge as bright and new beings. We bloom and we blossom again.

These sheep, these twelve ladies, are part of my landscape of the farm now and part of that blossoming. Like the mountaineer and writer Heinrich Harrer did in his seven-year stay in Tibet, I am coming to know my new, old home. When I think of Harrer, who fled to Tibet when he escaped a British prisoner-of-war camp in India at the outbreak of World War II, I reflect on what he found there: a version of himself hitherto unknown. I read his book *Seven Years in Tibet* one long grey winter after I'd returned to Ireland and felt, like him, that I was discovering a new land, a country unknown to me because I had left it so young to go out into the world.

The Tibet of Harrer's words is a now-lost world, for that land is no longer free. I, too, have been in the presence of the Dalai Lama and like Harrer have found myself a changed man for having met him. But perhaps I am not the first shepherd to meet the great holy persona, perhaps he too has a special place for the people of the earth. In Tibet they call him *kundun*, which means simply 'presence'. That feeling means much to me now. We can grow in presence. We can be our best selves in presence. The seven years have perhaps taught me that. If we meet the Buddha on the road, we should not kill him, as the old masters say, but rather embrace him in whatever form he takes because he has much still to teach us.

It is now late autumn, and the girls have been grazing the big reseeded field nearby for a few weeks now. Their bellies are growing fatter, full of new life. They must eat lots of grass to grow all this coming life. In the mornings, we bring them extra hay to help build up their bodies, and between the hay and the grass they are coming into something new, something sturdy and healthy. The blue Texel ram is with them now, though his job is long done. The twelve sheep are, we joke, his girlfriends and keep him company. He is a good beast and the father to many animals in the commercial flock that my parents own. I like him because he is part of the fabric of the farm. He has lived longer than many of our other rams. I like him, too, for he is not hard delivered, meaning that his children are not hard birthed, and they have been lucky in their births bringing live lambs into the world – that, above all, counts for some good in this endeavour. Luck is important in farming.

Seeing the sheep eat, I am happy. Let them grow and mature. Let them become mothers. When the grass runs out in this field, we will move them to the next paddock and the next. There is lots of grass for them. This green pasture is part of the creation story of this farm.

Home is something I have thought about a lot in the last seven years. My wife has spoken to me about how precious it is to be rooted in one place for generations. She is the daughter of Vietnamese refugees who fled their home at the end of the conflict that her people call the American war. They ended up in Australia, her birth country, on the decision of a coin flip.

For me, home was not always this way: for a decade I lived as a migrant in Australia and Canada. I was never rooted in a place for more than a year or two, moving houses and flats as tenancies expired or new love bloomed. To be a migrant, there is always the before and the after, the departure and the return. There is forever that fault line. I can still walk the streets of Sydney or Toronto in my mind, but they are no longer charged with connection, with meaning. Birchview, home, is the centre now. It has taken me years to find that sense of rootedness and it has not been easily won.

What is home? I have asked myself that question often in these years at home after all that time away. Maybe it is simply a feeling. I will forever be the man who left and forever be the man who returned. I am both things, like all migrants, but I have come to understand that that is OK. It is OK to live in one land and think of another. I think, in a way, these sheep are rooting me anew. I am, perhaps for the first time in my life, grounded in a landscape, and that feeling gives me a vision, a picture of life in new ways. I can give myself the permission to make a home in Birchview. That permission, I see now, was always within my power, but I did not see it until this moment.

The sheep make me feel a new-found sense of prosperity. Not material wealth, though that is nice, but something else, something more profound: it is, I think, something of the earth and placing oneself *in* the earth. To me, the rural world is home now. Cities feel different to me these days. They are still places of great excitement and potential – and indeed I like certain things that I can get in the urban world, certain speciality foods or clothes – but my nights in the cities are coming to a close. Maybe I have simply changed and no longer want the crowds. I yearn for trees and mountains, for contemplative scenes.

I feel more at ease in the deserts outside LA than in that great city, though I am in that huge urban place for work at least once a year. I like the quiet of the sands outside that city. I like to be in the echo of the wild nature of that place – it gives me peace. Once, in Joshua Tree, I camped out for an extra night just to ensure I did not have to spend a day in downtown LA, so unused to urbanity had I become after months on the roads and in the parks of rural America.

I never felt truly rooted in the cities in which I lived: the houses, the apartments were never mine – I was just passing through on my way to the next place, the next avenue or boulevard or adventure. I think I have taken the best things from the cities and brought them to my rural life. I am a product of the city and the country. I am a multitude of both. I celebrate both aspects within me.

I like this new relationship I have with land, with open space. I feel a connection to this rural world. It reminds me of what Narcisse Blood, a First Nations tribal woman and keeper of the thunder pipe bundle (that holy object of smoke pipes that is so precious to her people), said about her land: 'For us, relationship is our life – the relationship to the land, the relationship to the bundles, the relationship to the animals that are in them, the relationship to the cosmos ... Everything is about relationship.'

When I returned to Birchview seven years ago, I came back not richer but perhaps wiser. The cities had taught me many things. I had lived lives in them both good and bad and had learned a lot about life in the process. But in so many ways what stability I have built has come from this land and writing about it. That stability was hard won. John McGahern, a writer who lived in the next county and a man my father used to meet at the livestock marts of Leitrim, said that the local was the universal. In a sense, I have formed my writing life in this small universe of fields. I have re-rooted in the country, in the *terra sanctus*, or blessed ground.

As a place, Birchview is plain but its history is long. There were wars and rebellions fought on and around it; it has known chieftains and kings and soldiers; even the sheep and cows are the subjects of myths and legends that flow through these very fields. I have but to look out across them to see the trail of the Cattle Raid of Cooley, the founding epic of this land. Here the great bulls roamed and the war of a Celtic queen, Maeve, against Cú Chulainn was fought. I live, I suppose, in a palette of stories and I have taken it as my job to write them or interpret them anew, such as I am fit. Here, the Celtic world that made us is not dead: it has merely changed its form – there are still heroes and villains in these lands. The epics are written and rewritten anew each day. Where once a god did battle with elemental forces, now a man or woman tries to bend nature to their own will. The Dagda, the good god, was the father of agriculture and fertility and with his wand could kill or bring things to life. He may be gone but he can still speak to us of all the quiet mythos of this place. Maybe in some queer way the Dagda is in these sheep: with his magic he quietly guides me, even though his name has not been spoken in these fields for nearly 1,500 years. The legends, the myths, they surround me, they let me know that we can see the landscape truly for what it is: our birthright.

I have had meaningful and even profound experiences in other lands but they have always been examined through the prism of this place; they have been brought back here both physically and spiritually so they could be understood. So that the meaning could be found. Birchview has been a great threshold over which I have stepped to cross and recross the world, coming to understand it and myself in new ways. The great mountains of northern Spain or the desert plains of the American Southwest have been processed in my head and heart through these fields. As a nineteen-year-old, I left to go to Australia, where I compared the Great Dividing Range to the hills of our farm near

Clonfin. Things were measured against what I knew, what I had lived.

Souls and beauty are something I think of a lot when I think of this farm, this place, this home. There is a 'clumsy beauty', as Thomas Merton the Trappist monk put it, a soulful clumsy beauty to it. There are places in the land I know intimately. Places where I have wound wire and wood to make a fence. Gaps where I have planted gorse or holly. The fields, to another, might be static things, but to me, they are far from that; they are things of this living palette: a rock, a gate, a meadow, a stream.

Homes are, I suppose, clumsy by nature – the best kind of homes are, at least. They are not clean and ordered: they are waking, moving things that show us how we live and, if we pay enough attention, how *to* live. A field such as the Bottoms in Kilnacarrow, where the sheep will go later in the year, has the soul of wildness still in it. There amongst the rushes, which we think of as a weed, the wild earth battles with the tamed one. I like this idea, this struggle. The wild land where humans do not go so much is special; it regenerates into something different. It is as close as I can come to some African landscape, with its uninterrupted wild grass and trees, and only nature knows why it is this way. The mind of nature is a deep one, deeper than we shall know.

The monks, both Christian and Buddhist, sought the landscape of quietude in places such as this, far from the crowds and the towns. I am here to gain that too. I am here to grow like all the things around me. I will grow with these sheep, with their as-yet-unborn lambs. I will grow and the earth will leave its markings upon my face like it did to Narcisse's people, as it does to all the people of the true earth.

On the land now, I find myself more in my true nature. I am an authentic person on authentic ground. I write to understand it all but I farm to live and know. The outer world, the world of cafés and traffic jams outside the farm, could never truly replace this landscape for me; not now. It cannot understand the dance of life and the courtship of death we live with. It cannot beat the creation game, for there is too much newness to be had here in the now, in the universe of the real.

I will grow in the presence of home. I am ready to sing its song. It is a lesson I will never get tired of.

LESSON 3

Walking is good for the soul

'Walk as if you are kissing the earth with your feet'

Thích Nhất Hạnh

S ome lessons are best learned early in the morning, when the dew is still on the grass and the late-winter light, which is slow, is still dawning over the fields. I like to go and look at the girls early. It sets up the day.

It is December now and the sheep have grown fat. We will scan the girls soon to see if they are in lamb, for their thick fleeces prevent us being sure of it. I will not see lambs jumping in bellies, but I like to imagine it. I like to think of all the coming joy. There are, no doubt, twins and maybe triplets in there. My mother is confident they are all in lamb. Time will tell the true reality. The sheep will carry their unborn for 152 days. Some are a few days under, some are over, but the months do not lie. The months tell us that time is passing, that the great event is coming nearer.

The ram is now in a paddock of his own and he is happy eating grass and getting the occasional bucket of nuts. He is in good form and will last another season. We are happy about that fact. He is a good beast.

My walk to the sheep is now a kilometre or more from the farmyard, up by the river and to the new paddocks in the upper ground. They have eaten the grass of the lower fields out entirely and it must be saved and replenished for spring. In the distance, I can see Cairn Hill sprouting anew each day from the earth. At times, great clouds cover it, making the hill look like a mountain, and in that I take great delight.

The hill, known long ago as Sliabh Uillind after Uillead of the Red Spear, is an ancient space. It is made of Devonian sandstone and conglomerates. These rocks are found from the east coast of America through Ireland and Britain and on to Norway. But it is the Ordovician slate found in the northern parts of the county and on top of Cairn Hill for which I have the greatest love. This stone is ancient, some 485 million years old, and is the remnants of a former ocean floor. If the creation of the world took place in a day, a report on the geology of Longford says, the slate on Cairn Hill would have been formed at 9.28 p.m. What is special about this Ordovician rock is that it is also found at the highest point on earth, on top of Mount Everest. In this, our hill is a sister to the giant mountain of myth and legend. It is our Everest in Ireland, and in walking it, we can be our own mountaineers. Though 8,571 metres of height separates the two peaks, they are both no less sacred, both no less storied.

In a way, we are all of us linked by stones; they form the bedrock of our lives but also something more, something striking. I think of stones as our makers. The world was once their domain alone. The plants, the fungi, the animals – it all came later. The stones were here first. Stones are solid, yes, but like the limestone bed that encompasses so much of the farm, they are also porous, shapely and in no small way beautiful. The rain and the elements have made captivating shapes of these supposed dead things. When my wife and I travel to different countries, we like to collect a stone from each place. In my house, sandstone and granite sit together in a small bowl on my desk, my geological orchestra. They all ground me, and for that reason I love them.

A heron rises up from the water of the Soran river, disturbed by my footsteps. I watch him lope slowly across the sky up towards the upper ground. Herons have fished this river as long as we have been here. It seems through all these years that it has been the same heron all the time, fishing and flying, loping and banking, but in fact it has been the parents and children and children of children that have tilled this water, eating its goodness, savouring its health. They have been lifetimes in this place, and they are as much farmers as we.

I am, I suppose, engaging in a sort of meditation as I walk up the fields. In this space, I can see the land, the sheep, the heron and the story of this place. I see that I am freed from a busy mind to a mind of oneness. Thinking, Heidegger, the German philosopher, says, is a gift of being. I am glad of this gift.

As the river flows, I move across the puddles and the muck and hear the rhythm of my wellingtons as I walk. I like the sound of wet earth: it lets you know you are alive and to pay attention to what is beneath your feet. It is just as well that I like it, because this is hard ground; it can be unforgiving in rain or drought. On the underside of the soil, there is an unseen universe full of worms and clay, badger sets and rabbit warrens, stone and blue dauby clay that, to my eye, looks like the kind that potters and sculptors work. These are all part of the gifts of this land.

Past the big paddock, which we reclaimed from the rushes a few years ago, I take the right-hand turn and open the old rusty gate to the upper ground. I am coming to see my girls but, looking now, they are not in the big river field but further up in the back fields. I retrace my steps of other days, seeing a footprint in the earth, passing through water and muck and listening to the river as I go. The trout will soon come upstream in the strong flowing water to spawn and the minnows will, I imagine, tough out the fast-flowing water to stay in the outcrops and deep beds of the stream.

Moving through the grass of the field I feel it crushed under my boots. The grass grew well into the late autumn this year, and though it is not plentiful now, there is still grazing to be had and the sheep are the animals to do it. With their small but precise teeth, sheep can overgraze land, so we must watch out for that.

A rain starts to fall and I zip up my jacket. It's a light waterproof number that will keep me dry. The rain falls heavier and I walk to the ditch and stand for a time. I do not need to get soaked; there is no rush to see the sheep. I put my hands into my pockets to keep them warm, and in the action, I find something hard. I take it out and smile to myself. It is my scallop shell from the Camino, that great religious walk across the north of Spain. I travelled it a few months ago. It was a powerful experience and, in ways, a life-changing one because it led me right back here, of all places. I feel the shell in my hand and run my fingers across its ridges. The shell is the symbol of the Camino, and this is a special one from a special time, not least because it was gifted to me by Father Sean, my *anam cara* or 'soul friend' and local parish priest, before I left for my great walk. It is special as it reminds me there is something in us that needs to move. The Camino taught me that.

We have always sought the horizon, the never-ending skyline. From our earliest days in the Rift Valley, the womb of our species, we have moved and, in that movement, found our first beauties: a sunset, a mountain, a river, the sacred elements combining as the world was sung into existence before our feet. Our love of walking goes beyond the finding of food. The nomads, by way of example, point us best to this, from the Aboriginal tribes of the central deserts of Australia, about which Bruce Chatwin wrote so beautifully in *Songlines*, to the bushmen of Africa, who Chatwin said would name the contents of their territory forever in their wandering and thereby become 'natural poet[s]'. Writing on the subject of the nomad and walking, Chatwin said: 'Man's real home is not a house, but the road, and that life itself is a journey to be walked on foot.' In this I agree with him. To walk is to live.

Erling Kagge, the Norwegian explorer, says that your feet are your best friends because they tell you who you are. In this wisdom, I find a new understanding of my own feet. They tell me many things: that I am a wanderer coming to rest, that I am a farmer and perhaps, if I am lucky, that I am in commune with the ground.

I had thought about the Camino for a long time before I undertook that journey. I had been told that the road sends messages to the walker. That in northern Spain, when you walk you meet a new side of your soul. To walk the Camino is to be a pilgrim: one takes a journey to find a new horizon, as our forefathers did, and, if one is very lucky, to find a new way of being whole. I myself was looking for a connection, for a message about the future, or indeed from the future, about what I should do with my life next. I had lived many lives already, from being an investigative journalist in Australia to a filmmaker in Canada. I walked because I wanted a guiding light to the next part of my story.

There is no one single route on the Camino – rather, it is like a spider's web of various pilgrimage roads that fan out across Europe from Britain and France, even as far away as Poland. All the roads, however, eventually lead to the shrine of Saint James in Santiago de Compostela in Galicia.

Just as there are many different routes, there are, they say, many different Caminos: Caminos of love, Caminos of illness, Caminos of solitude and Caminos of healing. I started my Camino in Roncesvalles, in north-eastern Spain at the base of the Pyrenees, in an old monastery where three monks still live. I had an idea of what my walk would be about, but then the road, *the way*, had other ideas. When I walked, I was sure that my message was going to be that I should leave the farm and return to college to start divinity studies. Something had been telling me that for some time.

I felt, not for the first time, that I needed to engage more with my spiritual side. In Canada, seven years before, I had contemplated becoming a priest, and later still a Buddhist monk. I put those ideas aside and thought that maybe I could answer my longing through writing, but the spirit is strong stuff: it comes to the surface sooner or later. I read the works of religious leaders, I meditated with Thích Nhất Hạnh, I began to go back to Mass. I visited temples and shrines and churches in different countries and I thought that perhaps coming closer to God, to the creator spirit, might allow me to understand myself better. That, in short, if I devoted myself to divinity studies the message might come to me. That perhaps in all those books of learning I could find a theological guide on how to live. Maybe that's what all religious lives are, attempts at understanding how to live. When I found myself on one of the most sacred pilgrimages in the world, it made sense that I would continue to carry out my work into the divine. But it was not to be.

Aislings are an important part of life to me. An old Irish-language word that translates to a dream or vision poem, aislings can also work great wonders in our lives today. They are important because they let us know we are alive: to have a vision is to want to continue. To the people of earlier generations the aislings were a sign of our freedom from oppression, but perhaps in today's world they can be a sign of personal liberation and our destiny. Destiny comes through the feet, through moving toward the goal. Aislings matter, dreams matter.

In Roncesvalles, in the old museum of the monastery where the wealth of the Kingdom of Navarre is held, I received my first aisling in a painting of the Holy Family with little Saint John. In the *Sagrada Familia y San Juanito*, Saint John is portrayed as a young child holding a tiny lamb in his hand. As with all signs, when we first see them, we think nothing of them, but there, on my first day of the Camino, the *cordera*, the lamb, came into my life.

The land of the Kingdom of Navarre is lush and green, and even going so far from home, I felt that I had not left, because things were alive in the way they are on the farm. When I walked in the land and countryside of Navarre, I captured the same feeling and the quietness: there was no noise but the crunch of my feet on the gravel.

For a time, our guide was a writer, Fran Contreras, who had published a book about the magic of the

Camino after walking the route twelve times. Talking one day in basic Spanish so I could understand, he told me that the pilgrim route was not about religion for him, that, in fact, his gods were the moon, the stars and the mother sun. That he was, in short, a proud pagan, that the road had taught him to see the symbols of life everywhere, not just in churches and holy spaces.

To be a pilgrim is to ask the question: why am I here? I think that we must all ask ourselves that question, pilgrim or not. We must discern what our lives are about and what we are walking towards. In ways, I have been discerning my life for so long now, but it is good to seek, good to delve into the mythic mind and the mythic land.

The pilgrims of the Camino come from all over the world. One evening we met a group of German, Irish and American walkers. They had banded together to form a walking family. Over dinner one of them, a man from County Down in Northern Ireland, explained to me that it had been on his list to do this walk and it was now or never. The message, the meaning of the Camino, had not come to him but he hoped it would before he reached Santiago. He explained to me his idea that the walk was the vehicle for the message: on the pilgrim road, you have to slow down. You have to come to the realization that you are on, or rather in, another timeframe.

It was in Puente la Reina, a small town in Navarre, where I saw the second of the signs that told me that I myself was meant to be on this road at this time. One evening, on a quiet street after a hard day's walking, I heard bells ring out loud and clear. I looked around a corner for the source of the noise to be greeted by the sight of a young shepherd and his hundred sheep or more. The shepherd was no older than me. I knew then that the Camino was telling me something. That the sign I was being given was not to go back to college but to return to the land. That, in short, my way through the Camino was the way of the sheep and that my walk would not be over mountains but through fields, our fields.

Looking at that young shepherd, I thought of Paulo Coelho and his shepherd boy, Santiago, in search of his destiny in the writer's most celebrated book, *The Alchemist*. Like Santiago, I was wandering the world only to come to the realization that my future was at home all along.

Coelho has been a guide of sorts for me, a writer with heart and bravery who believes in magic. I have that self-same vein in me: that want for magic, for life to speak to us in strange and beautiful ways. Coelho risked so much to find out the meaning of life, from his days as a patient in a psychiatric hospital (having been put there as a teenager by his parents) to practising black magic, yet as he says himself, the meaning of life is the meaning we give to it. But we must all find that lesson ourselves. In ways, I think, we must all have our revelation moment, as I did on the Camino.

Sometimes it takes magic to have that insight. Coelho says that following your personal legend is your blessing for following the path God has chosen for you here on earth. He writes: 'Whenever a man does that which gives him enthusiasm, he is following his Legend.' The legend is our mission in life, to become what we were destined to be. That gives me great heart. I want to follow my path. I have been trying through all the obstacles and failures. Maybe this is the path, the blessing God has laid out for me. Coelho has reached so many with his words; maybe mine can help some people too.

Some journeys take longer than others to complete and my walk on the Camino was not over that evening: rather, I passed through the lands of Spain collecting more sheep as I went. In the cathedral of Santo Domingo de la Calzada, in Rioja, there were lambs on the ornate altar created by the mediaeval artist Damián Forment. In Castilla y Leon, I was greeted by a road sign with a black sheep on it. Why the sign was a *black* sheep I will never know, but in my state of mind that evening after a day of psyche searching in the decaying churches of the *mezeta*, the forgotten plains of Spain, I took it as an emblem of all the black sheep of the world, all the outsiders, myself included.

When I think of those towns and villages now, I think of the warm evening air, of the sight of fellow walkers on the roads. They were all seekers. Perhaps the villagers of these towns are used to seekers. It was a new feeling for me amongst the sandstone bridges and the quiet venerable churches. When I looked, the signs were all there. Maybe divining the secret of life is all about symbols. as Coelho rightly says.

When at last we came to Green Spain, to the region of Galicia, I sighed a breath of relief to be home with my Celtic brothers. There, the sheep were strong and thick-set, like the sheep of Ireland. The farms were placed within the hills and mountains, not seeking to dominate the land but, rather, to coexist with it. It was the Celtic way.

On one of my last days, I met a Jewish aeronautical engineer who had worked for NASA launching rockets and satellites into space and who now, upon retiring, was wondering what to do with her life. I told her then what I had come to know: that we bring our questions with us on the Camino but we bring the answers too. The walk is a way to open that answer. For her, what I said came as a revelation; to me, it made sense. It is, I see now, the road, that unending earthen path, that gives us not the conclusion to our problems but the time to think on them.

On one of my last days on the Camino I experienced a night of severe pain in my chest. It was intense and lasting and I thought that I was going to have a heart attack. I truly did. I remember saying a prayer that night before I went to sleep, thinking that I might not wake up again in the morning. I even phoned my wife and said my goodbyes. Why I did not send for a doctor, I am not sure. I was OK with my life, I thought. I was happy with all I had done. The road had let me see that there had been magic not just in the walk but in my whole journey to that point. I was happy with all the ups and downs and saw that even the darkest moments, and there had been plenty, had helped shape me into the being I was supposed to be. Whatever the issue had been – I will never know – I awoke the next day alive and thankful. I could see beautiful nature all around me. Maybe in some sense part of me did die in that hotel room so that another could be birthed. It was my crossroads moment of life.

As I walked the last five kilometres into Santiago de

Compostela after a gruelling, long trip, I was greeted by the sight of a field full of lambs running towards me. The sheep were guiding me home, I think now. They were as good as any satellite or space shuttle from NASA. The message was present. Maybe it was magical thinking. Maybe it was religious. When I entered the tomb of Saint James, I said a quiet prayer for my wife and my family. I did not ask about my future, as it seemed to me that question had been answered. In that little passage at the back of the cathedral life had found me. The aislings, the lessons of the road, had come to me. I had awakened. I had achieved continuance.

There in Santiago at the end of my journey I became a vision poem for myself. We can, I can see now, all become that poem; all who are brave enough to walk can become the people they want to be. My next step was to be a shepherd – in ways, these sheep were always calling me. The Camino showed me that above all else.

The heavy rain has slowed and I put the scallop shell back in my pocket. It now comes with me on the farm for this place, too, is holy ground and my walks on it are just as important. As I walk up the field and cross the gate, I see the hoggets. They don't run away from me as they did at the start, months ago. They are used to seeing my form each day. Soon, it will be time to move the sheep inside. We do not want to risk – I do not want to risk – lambs being born outside in the cold weather. As they are first-time mothers, there is a danger that they may have difficulties lambing. I want to be there for them and for myself.

I am excited to think the lambs will come, that they will run in the spring grass, but the winter must come first. The season must be got through.

I walk gently through the flock and talk to them as I go. I tell them they are good animals. I do not know if they understand me. But in patterns beyond speech, beyond the language of humanity, I hope there is comprehension. Animals can, I think, feel the world around them far better than us.

I never tire of this walk: each day there is a new thought to be had and new sheep to be met. They are not signs now but the real embodiment of my own flock. They teach me like the painted sheep of Spain I saw. Each day is a new learning. It's a way that I can keep the Camino going, as Father Sean calls it. There is a *hesychasm* in the walk. The literal meaning of the word is an inner stillness, a divine quietness, so as to come to a meeting with the creator. I met him on the Camino. I meet him now with the twelve sheep.

Walking is good for the soul.

LESSON 4

In waiting,
we grow

ime has moved on and so have we. December has passed and the sheep have been moved inside, into the sheep shed, because the lambing season is nearly at hand. All the girls are here now, waiting. The twelve hoggets, my twelve sheep, will soon be mothers. They are ready.

Waiting is part of the mystery of farming. Waiting for grass to grow, waiting for crops to ripen, waiting for new life.

In the quiet of the shed I bring out holy water and throw it over the girls so that they will be safe in the coming weeks, so that the lambs will come true and strong. It is a ritual I have picked up from my mother. It is a simple act but an act of great significance, none-theless. I do it to ask Saint Francis, the patron saint of animals (and our farm), for help but also in no small way to ask the whole universe of animal gods.

In the past, the sheep was worshipped not just as a provider but as a god, a deity that conferred more than we can know in this day. When the first peoples domesticated the mouflon sheep in Mesopotamia, they began a courtship that has lasted eleven thousand years. In the lifetimes of sheep, it is long ago since their mothers came down from the mountains to join the new family of humanity. Skulls of rams were placed in the houses of the people of ancient Turkey, showing their importance even then. In Egypt, the ram-headed god Amun was known as the god of fertility. As far away as China, the animals were seen as the bringers of creativity. Indeed, Pliny the Elder wrote of them as an animal we owe great thanks to in his tome the *Natural History*.

Sheep have always been part of the story of this island. Our relationship with them has been both a spiritual and a practical one. First brought to Ireland by the Neolithic people, these early arrivals were said to resemble the modern-day Soay sheep of St Kilda, a small hardy breed that hold onto many of the old traits, such as horns and colouring. The ancient legends of this land make mention of our great ancestor sheep, Crib, who, along with a boar, was an accompanier of the goddess Brigid, a member of the Tuatha Dé Dannan and daughter of the good god Dagda. Brigid was associated with wisdom and poetry as well as domesticated animals. Back then, to have an animal not only symbolized a richness of physical property but also marked a spiritual link to all our brother and sister animals. In ancient times, when we looked at a ewe or a ram, we saw not just an animal but a kindred spirit.

According to the academic Fergus Kelly, texts from Old and Middle Irish detail our long and storied history with sheep in this nation. For instance, they were not considered beasts of the hill country then and were grazed on the greens and lawns and plains of the land. Also, in the old law texts of ancient Ireland, sheep were protected and honoured. Sheep, the old records recount, needed to be protected after dark from wolves and wild dogs.

And we humans were not shepherds then, rather we were *augaire* or 'sheep callers', for those who called the sheep to them were their masters. In the old days, the augaire led their sheep rather than followed them, as we do in the modern day. It was a time of a closer bond between man and beast, though it should be noted that the occupation of shepherd was considered of low status in the old Ireland, something akin to a cowherd, as Kelly relates it. It is strange that sheep could be of the gods and goddesses but the carer was not given as much thought.

The legend of the Golden Fleece is rightly celebrated around the world but our own legends with Crib and the coming of the sheep to this landscape of poetry have a place in the world's culture too. These animals are the holders of a lifetime of fleeces, yes, but also stories. Crib, even removed from us now for all these centuries, can teach us still.

The shepherds of yesteryear were great people for waiting. They knew that everything had its time, its place in the seasons of the year. They, too, began in March and April with lambing and did not let rams to their ewes until October and November. It is a cycle we shepherds follow still.

One of the big things, one of the last things, we understand is ourselves. In waiting, I have found an understanding of my relationship with the earth. In waiting, I have found myself to be part of the unbroken chain from the first farmers who combined the natural and spiritual world. The first men, the first women, they made this world, these fields, this culture, this way of being. They, too, waited just as I am now doing.

As human beings, I think we find ourselves in waiting. In the mood between the great events. In the gaps between the temporal markers. As E.M. Forster said, 'We must be willing to let go of the life we have planned, so as to have the life that is waiting for us.' I think I know what he means now. I never thought that I would be a shepherd, yet I find myself wholly invested in this work, in this endeavour. I was once an investigative journalist. I was once a filmmaker. I did not seek the knowledge of land or animals but the secrets of people. Now, I think that the nature of the endeavours was the same: I was always looking for truth. At first, those truths came from a leaked file or a distressed refugee; now they come from a quiet sheep. I see now it was all a form of mining, digging down for the realness.

I understand myself better now as I wait. I understand the land better in its own song, the song of the sacred. I have, I suppose, come back to the land to find peace in a weary world.

In the dark nights to come, I will wait in the sheep shed. With a cup of coffee and a long book, I will wait for life to bring life. This season I will read the work of my old friend, David Malouf. He is thousands of miles away in Australia but his words still teach me, and though we do not talk as much any more, I am still learning from this master of the word.

I will pray when the lambs do not come right and when they do.

There was a time when I did not pray, when I did not give thanks, but over the last seven years, I have been on a spiritual journey, a journey that has seen me question my faith in order to understand my *fate*. It seems to me now that faith and fate are connected. All spiritual lives are concerned with fates. We are all trying to discern what the future holds for us. When I was young, I had a child's view of faith but that grew and evolved as I began to ask deeper questions. On my travels around the world, I talked to faith healers and witch doctors, sages and priests, and they all spoke to me of the need to be connected. That it is connection that brings us peace. My seven-year story in this land has been about that, too, I think. We must all be seekers in this life and not be afraid to take the hard roads to gain a form of understanding.

In the time that I did not pray, I was not asking myself about my life; when I took to prayer again, I started the conversation that has brought me here. As beings, I think we can find ourselves in prayer. We can understand ourselves better in the sacred song of the earth. Our days are so dear to us, though we do not count them as dearly as we should. Hesse rightly calls them 'precious' and notes that we 'gladly see them going', but these precious days do not just shuffle by: they race past us so quickly we are unable to savour them. Is that part of fate? My fate? I think so, but I have woken up to this race myself before it has all slipped by and for that I am glad.

I walk into the shed now and marvel at the girls. Waiting has brought us this far. The scanning man came a few days ago to tell us how many lambs to expect and we have found all the girls to be in lamb, all ready to take up the task. He marked their backs with spray paint to tell us the numbers to expect. We know the numbers but we can only guess at the intentions of life and nature. So now, like a soldier along the black-watch wall, I wait.

In the lambing season, we can be the shepherds we were always meant to be. In past seasons, lambs taught me that I could be a shepherd to them, and I think now they teach me I can be a shepherd to myself. This time of waiting, this blank patch, has become a great lesson to me. I walk out several times a day to see if there are swollen udders, to see if there is the birthing slime of broken waters. But so far, my checks have been unrewarded. I come and find nothing, only the gentle faces of the creatures looking at me.

Waiting is about patience, and I am learning that now. I cannot rush nature: it moves at its own flow, with its own grace. Nothing can be rushed into existence. Like the woodturner taking the bog oak from the peaty moss, there must be patience until the sculpture emerges from the wood that has been waiting for thousands of years to become its true and finished form. In this context, what are a few more days, a few more weeks, in the grand scheme of life?

I will use this waiting to be a teacher, to begin a true relationship with myself.

Carlo Rovelli, the physicist, says that to be an individual is a process. That we are complex and tightly integrated. I am interested in that word *integrated*. I think Rovelli means we are the linking of the physical and the mental and maybe, though he has not said it, the spiritual too. Thus we are physical but also so much more. To be an individual takes an effort all our days to integrate, not just ourselves in all our various forms, be it bodily or spiritually, but also to encompass life and time. To make sense of it all. In the prayer of life, the good and the bad must be brought inside us.

While I wait on the farm, I take the lessons of life, the hard won and the easily earned, and I give thanks for them all. Waiting is my home. Waiting is part of life.

We must keep wonder in our minds

t's the afternoon, and I have spotted her, the first hogget to go into labour. It is not sudden or startling, it is natural and plain, and the bag of life is hanging from her. In the early days when we first got into looking after sheep, I would have rushed to take the lamb from her, but now I know better. I know that I must wait for her to open fully, to dilate, as a vet would say.

She is in the upper shed with the rest of the ewes in the loose pen where they have been for a time. My first job must be to move her to a pen of her own. I take fresh straw from one of the round bales and make ready the lambing pen. I put together the first few pens a few weeks ago with the sheep gates Da bought from a travelling salesman years ago. There are six pens set up. Enough to do us for now. There should not be six ewes lambing at once, but when the humour gets into them to lamb, it can spread. Where there's one, there are often two mothers.

I always set up the pens myself. It is my ritual. I make a note of the appointed day to build them. We cleaned out the shed at the start of the season with the digger and the place was clean and ready to be used. We had metal feeders for hay and plenty of green plastic buckets for water and feed. The whole job took an hour or two after I first disinfected the ground of the shed to kill any of the badness that might seep into a young lamb or a sick ewe. When the shed is set up for lambing, it looks like it has found its true purpose. The shed has taken on a feminine role, and it has become a mechanical midwife, a holder of futures and, perhaps, in no small part too, dreams.

The ewe is smart and knows that something is up with her. Together, she and I now walk out of the upper part of the shed and I show her to her quarters. The straw is fresh and rustles under her. After she runs inside, I close the gate. She is secure now and I breathe a small sigh of relief. That is job number one. I fill a bucket of water next and place it inside the pen. I will need some to feed the ewe and some to spill onto the lamb when it is born to bring the shock of life into it. Next, I take the lambing gel and a white lambing rope my father got last year. These, along with my hands, will be my tools. I look at my hands then to ensure they are clean because they will be inside the ewe. I have cut my nails short in preparation for this day: the lining of the sheep's uterus is thin and I do not want to cause any damage. I take off my wedding ring then and my good-luck Claddagh ring. I will not put them back on until the season is over. I stow them in my back pocket. I am clean and ready and it's now or never.

Between the moving and the preparation, the ewe has had ten or fifteen minutes and I judge it now to be the right time to take a look. Da is away and so I will hold her and check her myself. It is not so big a job, as she is a quiet animal and I think she will not mind. I wet my hands with the blue lambing gel and lather myself up. I grip the ewe's back and force her to the ground, and she lies out flat and her chest rises up and down and she looks at me. 'There, there, girl,' I say. There is no worry in me because I have done this act so many times that it is natural. I push my hand inside her cervix and break the bag and birth fluid runs out on my arm and then I feel the feet with my fingers; they are my second eyes.

When we lambed our first ewe seven years ago, we knew nothing of this life. My father and I were novices then. When we found a tangle of legs that day, we did wrong and pulled hard with ropes until blood came from the cervix, and we knew we had done bad.

Failure taught me that day, and I went on the internet and got a vet's instruction book on birthing lambs. The document was full of illustrations of all the different ways that lambs can come: upside down, backwards, breech births and the tangle of legs, which I know now to be the sign of twins. I printed out the book and it lived in this shed for that first season until it was all in my head and I had learned it off by heart. I came to know the world of lambs and ewes and birth and death too. There, in that book, I found the key to the book of beginnings and, in no small way, life itself.

In searching for the lamb, I probe inside the ewe. There are two legs and I smile and I push a little deeper and find the head and it is upright and coming right and I sigh, for all is good and correct. I pull the first leg out and feel the slight pop as the leg straightens and comes out of the mother. I take the second leg then and follow the same actions and all is as it should be. I pull then, slowly but firmly downward, and the ewe kicks a little at this sign of pressure but, good girl that she is, she stays down on the ground.

'I have you,' I say. I pull stronger now and the plop of the head comes and it is a big head. The lamb has its father's head and I admire it for the briefest of seconds. I pull still more and the rest of its body comes out, fluid and easy, and with a damp thud, it is on the ground. I clear the birth film from round its nostrils and then taking the bucket of water pour a small amount on its head, and it shakes now with the water of life and starts moving. There is a wonder in the moment, wonder in the action, and I know now that in the wonder I can discover myself. I am, after all this, my own true shepherd. The birth has tested me, has pushed me to see what I am made of, to see if I can rightly call myself a farmer. It is a good feeling looking at this little lamb as it comes to life. I check its sex and see that it is a ram and I am glad of his arrival. Everything else, all the other roads of life, seem unimportant in this moment. Here in this little shed all the world seems to shine out to me in its beautiful nature. It is why, I suppose, the act of birth transfixes us all and has done throughout the millennia. We see the newness in everything in that moment in an old world.

Waking out of my joy I bend low and listen to the lamb's breathing and hear a small rattle in his chest, and so I take him by the back legs and stand up and swing him back and forth gently. The action dispels any fluid from his lungs, and when I put him down again, I come in close and cock my ear to listen and hear that the rattle is gone.

The ewe is still lying down and I break the seal of her tits and the milk comes out after the wax and I know that all will be right. In the beginning, I used to put every lamb to its mother's udder, placing their little mouths around the tits to ensure they had suckled, but now we know that nature does that for us, that the lamb will get the sauce of hunger and investigate and probe and eventually find the tit and suck on the goodness. It is all in his head and soul: he does not need me. A lamb wants nothing more than to be a lamb.

This is not my first birth but he is *my* first lamb and there is an importance in that, a wonder in that.

I give the ewe a push and tell her that she must be a mother now. She wakes from her labour and stands on to her four feet, turns and smells the lamb and nibbles at him, and this little Bethlehem scene is whole and new.

We are here to experience wonder. That is what we have come into this life for, I think. From the birth of a lamb to a sunset over the fields. To not live a life of wonder is to not live at all. Becoming part of the great symphony of life such as bringing these lambs into the world is a calling. I have been invited to take part. I like that feeling; it seems now to make more sense to me with every passing day.

This lamb, this new birth, makes me think of my recent, long voyage in search of wonder. In the dark of a northern-European year, sometimes wonder does not come into our lives at all and we must do things to remind us of the importance of that notion. Sometimes we must journey into the unknown to find it again. I carried out my journey into wonder months ago, on a solitary trip to an ancient place, but here, now, looking at this lamb, its real meaning has finally come.

It was the end of the equinox and the turning of the seasons when I left on my voyage into the Celtic valley of death. I was still in my space of emptiness, still in the days of my spiritual vacancy, and I was looking in an old land to find something new. I drove that day on a quiet road. The Covid pandemic was still in full effect, and though we could now move in that space, people were afraid to wander, a limitation that was against our nature. I had decided to visit the ancient site of Brú na Bóinne in County Meath in the east of the country. The land in this place is rich and fertile; the farmers raise cattle and they quickly fatten on the grass of its great plains. In the fields, I saw cows with calves at foot lazing in the hot sun.

Brú na Bóinne, or the Palace of the Boyne, sits at a bend in the river of the same name. It is a valley of mighty Neolithic tombs built before the very pyramids themselves. The goddess Boann gave her name to this place because she was its maker, so the story goes. The river, full of trout and salmon, has a magic to it too: it was here that the Salmon of Knowledge, the being who held the wisdom of the world, was eaten by Fionn mac Cumhaill. Fionn was never meant to know the knowledge of the world: he was only a servant to Finn Éces, his master, who had spent seven years trying to catch the animal.

When Finn did catch the fish, he instructed Fionn to cook the animal for him but not to eat it. The boy did as his master said, but as he turned the big fish in the pan, he burned his thumb upon the thin layer of fat that all salmon carry. Sucking his thumb in hurt, he extracted the power and mastery the salmon held. He then knew the wisdom of the world. Realizing this, Finn instructed his servant to eat the rest of the fish and so passed the Salmon of Knowledge to the boy who would be king and leader of the Fianna, the legendary band of warriors.

I have always felt an affinity with this land. My father knows this country because he has a friend who farms in the valley. The thought crossed my mind of connecting with him when my job was done and finding out what it was like to live in this place.

Newgrange, the mother of the three tombs that occupy Brú na Bóinne, is a 200,000-tonne passage tomb. Many years ago I entered its inner chamber, designed to capture the light of the winter solstice from the roof-box and tell us that we have survived another year. There is, in this place, a vision of life. It is not just the light that illuminates: I think in some way it is our very beings that light up.

The tomb might well be an ancient thing but there is something new about it. Its light, so precious and spare, reminds me of the 'corners of light', as the Dutch writer Henry Nouwen said, in my own life. 'Corners of light'. I like that phrase: it tells us that all dark places can be reached and turned from the negative to the positive. That is something pure and real.

Long before the site was excavated, before the roof-box was discovered, the locals knew of its power. They had retained its memory in their collective mind, and though no one living had seen the light in the chamber, the legend still held down through the millennia. When the archaeologist Michael J. Kelly began his excavations of Newgrange in 1961, the site was in very poor condition and all the treasures that it held were not fully known. Kelly's work, over fourteen years, would restore the place to its former glory. A mighty wall of white quartz was built to support the front of the structure and the many kerbstones of the tomb were restored.

The front kerbstone was what I had come to see on my most recent trip. Upon the stone there are carved markings which speak to another form of communication. The intricate swirls and lines were deliberately inscribed by skilled hands, using tools to pick out the designs upon the rock face. They were the earliest marks of wonder I have ever seen, the first marks which communicated a world that was at once my own and another.

I had seen markings like these in other parts of the world, from the American Southwest to Australia. Walking around the mound, I stopped and looked at the designs and wondered about the world they had been trying to replicate. What were their worries then? What were their fears?

I got talking to one of the guides, discussing the link between Newgrange and the wider world and what all these designs meant. 'We can only guess,' the man said plainly. I told him about my time in central Australia and the traditional paintings I'd seen by Aboriginal people, who used designs and motifs like this. For instance, the concentric circle in the middle of a stone represented a campsite or a waterhole. Interested, he told me that Aboriginal artists had visited the site a decade before and that they had been fascinated by the signs and symbols.

'Perhaps they could understand them,' I said. 'They are, after all, the oldest continuous culture on earth. They may not have lost the link with the original people.'

We were silent then for a time and my mind drifted back to the desert and the painted aerial maps that represented Aboriginal dreaming stories. Baiame, the great father of some of the nations of Aboriginal people, was said to have come from the heavens and created the land with all its features. Baiame and other figures such as the Seven Sisters also created the dreaming stories of nations. The people here on earth created paintings to depict some of those creation stories. Some even have songs to go with them. It is a rich spiritual and artistic tradition that many have tried to understand. Maybe the key to our forebears was hidden in plain sight on the other side of the world. Maybe an Aboriginal painter such as the great Rover Thomas could have told us what our people were doing and what they meant by these markings. The guide and I looked at the stones again and smiled.

Moving around that ancient site, walking over that ancient ground, I took in the creations by the hands of my forebears and all the work that has gone before me. Wonder built these structures. Sheer wonder for the world: from the question of the afterlife to the very real birth and culture of the everyday. Wonder is a powerful thing: it can accomplish so much, such as the building of these ancient tombs.

There was wonder, yes, but try as I could, I could not fully soak it all in, could not feel the connection that my makers had felt. It was a magical place, yes, it was a powerful place, but it did not stir in me the epiphany or revelation that I'd sought. I took the long drive home, questioning if I had missed something. Had I not been open enough?

Looking now, in the sheep shed, at this simple lamb, this living epiphany before me, I think that perhaps wonder was here in front of me all along. All I had to do was find it.

In another world, I would have, or could have, lived another life. My learning, though, is in this soil, with these sheep, in this shed. We must keep wonder in our minds, yes, but we must never stop ourselves from seeing it in the world before our very eyes. In the cathedrals of life is our best hope.

That equinox trip was, I think, a stage on the long road back to here, and its real message has only now hit me. We do not need to catch the salmon to gain its wisdom. It is within us all. We can all be fishermen of life, hooking its lessons as we cast out again and again into the great unfolding drama.

I get up now off my knees and look at the scene. The ewe stands over her panting little babe. He is exhausted from coming into this world but he will rise soon and then he will look for the goodness of her milk. I thank God quietly for the lamb and the gift of it all. That is one birth done, one more teaching delivered.

LESSON 6

Labour is unavoidable

The days are busy at the moment, with sheep in the maternity ward, their sisters who have not yet lambed in the creep area, which is what we call the straw-filled pens, and our other animals – the cows waiting to calve. The Buddhists say that now is the most important time in life. But on a farm, the past can be just as important as the future. They each have their place on the land.

Taking the nuts from the bag, I fill a bucket half-full with them and then add half a bucket of oats. The oats, my father tells me, give the lambs in utero vigour and the nuts give the mothers protein. I pour the mixture into the trough of the sheep who have not yet lambed. The ewes jostle and push one another to come to the feeding barrier. They are greedy and puck one another to get to the best point for eating. I tell them that they needn't worry and that there is food enough for all but they don't listen to me.

Next, I take a square bale of hay in my hand and cut the two pieces of twine holding it together with my knife. I take a sheaf of hay, like a chapter from a book, and give it to the girls in the pens. The hay is an act of thought and time. It began as a glimmer for this moment when the grass was still young last summer, and in that glimmer, we let the grass grow and, when the time was right, we cut and rowed and shook out the green strands until they were dry and golden. Hay is part of the simple rhythm of life and yet it is a brave act. It is brave because we are making a pact with nature that we will be around to use it. I admire the old men of the parish who still make their square bales of hay each summer; it's not just done out of necessity or tradition: it is their earthen testament to the world that they want to be around in the future to use it. Perhaps that is why my father calls hay romantic. It says so much in its simple form.

I take sheaf after sheaf of hay and place it down in front of the sheep and, eventually, the din of the morning calls quieten as the animals take to their breakfast. Eventually, all the sheep have hay in front of them. I pour fresh buckets of water for each of the mothers then. Some don't have the sense and shit into their own water, and so I check each of the drinking buckets to make sure the water is clean. Sheep can be slow this way but the cows are no better and I've seen them defecate into their drinkers too. The shit of the sheep floats on top of the water and is easily seen. I scoop it out or pour it out. Shit is something I deal with every day. You can, I think, find delight in shit, in the dirt of the world – you just have to get used to it.

The work takes me an hour or so but the time moves quickly. Labour is a part of this craft of farming. There are times when the farm needs more hands, but in modern Ireland, those hands are not always to be had and, besides, could not be afforded. Your family is your labour force, but then it has always been this way.

This labour I do out of joy. It is a necessity to help my family but I am not trapped by it, not bound by it. I can take a day off if needs be and I can rest, for the family is a team: we are there for each other.

There are others, however, for whom the work is not luxury. It is not done for joy but for survival.

I feel that I have a deep connection with migrants and their labours, not least because I was a migrant myself for many years. There are so many kinds of migrant: those who seek refuge from war and strife are one but there are also those who have come for a better job and life, to make a fresh start for themselves. In some cases there are those who are both. To be a migrant at times is to be a walker in two worlds.

Before the pandemic, I crossed the southern states of the USA meeting migrant farmworkers for a research project I called *The Broken Harvest*. The project has now been consigned to the top shelf due to Covid but it has been with me ever since. The names of the migrants I met cannot leave my mind: El Diablo, Carlos, Tierso, Lorena, Blanca. All of them are part of the story of the labour of the earth. In their hands, their earth-laden hands, I see another picture of farming and work. On my phone and handheld camera, I have pictures of all these people that I met and the conditions I met them in. They are all of them with me now in this shed. They are part of my walk upon the earth forever more. There is one picture that I look at often, of a woman. Her name is Maria Gonzales.

Maria doesn't know that I have often looked at this picture, that I have wondered about her ever since we met. She is on the other side of the world, unaware of me and my actions. I still have a way to contact her but I wonder if she would remember me as I remember her. It has been three years since we met. She will never see our farm, that I know. She cannot see this dimly lit sheep shed where I work now.

A picture is a testament to a moment in time: like a painting, it is static, as the art critic John Berger says. Pictures don't change but we can, like the seasons.

Maria Gonzales

Every time I look at Maria, I have a new understanding of the world she has lived through, the world she told me about. Sometimes I am happy when I think of her, sometimes I am angry. When I see Maria, I think of her labour. What has been given and what has been taken. When we met in El Paso, Texas, in January in 2020, over a communal Mexican meal of molé she told me a story. It was her story, though it belonged to others, too, but they can no longer tell it.

Maria came from southern Mexico in 1992 and became a farmworker. She came to America to find a better life because, as she told me that day, she believed in the American dream, that a person could make a better life for themselves in that country and that hard work was the way to do it. She worked in the huge fruit and vegetable fields in Arizona for many years, picking melons and lettuce, but the work was hard and she told me that they worked in searing temperatures of over forty degrees with no water breaks.

When we walk into the local fruit and vegetable market, we never think of all the hands that it took to get the produce there. We don't think that eating, that so very personal of acts, is, in fact, a shared experience. There is a ghost population sitting down to the table with us at every meal. The harvester, the picker, the washer, the delivery driver. We discount them or, rather, never even factor them in. I love tomatoes but I do not know how many hands touch them in the greenhouses of Spain before they come to me here on the farm.

After a few years working in Arizona, Maria had enough of that hot state. Her mind was made up, not by the baking temperatures in the fields but the fact that she was repeatedly not paid her full wages. They were docked or not counted correctly or stolen from her, and so she and her husband decided to start the American dream again in Washington state in the far north-west of the country.

There, she worked on dairy farms, where she milked 3,300 cows during her eight-hour work shifts. Sometimes, people did not turn up for their shifts and she would have to milk another 3,300 cows, working a double shift, a total of sixteen hours per day. I have never seen that many cows in my life, nor milked more than one cow at a time. I am not sure how one would go about that work. At most, there are a hundred or so animals here in winter and that is a lot for us. It is more labour than the farm can take sometimes. If we were to own 6,000 cows, we would farm the entire parish, maybe half the county, but then our neighbours would have no livelihoods. Family farming is the backbone of Irish agriculture; our small scale farms are what helps keep our rural communities alive. Things are bigger in Washington state, yes, but to have such scale here would destroy this rural world of ours.

Maria worked at her new job well and liked it. She milked the cows and cleaned the sheds, her husband by her side, and though there were a few accidents and one occasion where she was nearly crushed by the cows, she was OK. But then she was sexually assaulted in front of her husband, and when she complained to the owners she was fired because they did not want the hassle of her case. But that too was surmounted and they started again at another farm in the state in

2015. This was the year she would remember out of all the others.

When you get older, you don't get surprised as often by the world – you have seen enough to know that this is the way of things. But when I look at this picture of Maria, I have the distinct memory of the shock I felt upon hearing her story. It has created images, spectres, and in no small way haunts my mind. The conjured images have stayed with me, refusing to go away. They live in my imagination and my soul now in all their horror and all their truth. That truth, though, is part of the story of what makes Maria the woman she is.

His name was Randy. He was twenty-seven years old and he suffocated to death in a pile of shit on the farm. There is no other way to say this and those are the words Maria used to describe it to me. That the shit got into his nose and mouth and ears and eyes and that when he breathed in, he breathed in the cow shit.

Randy died because the brakes on the front loader didn't work. Maria said the employer wouldn't fix them. Randy was working the night shift on the dairy farm while his partner and children slept nearby and he went to use the front loader to feed the cattle and, in the dark, he drove into an open shit pit, which had no warning signs around it, and because the brakes of the machine had not been maintained, the vehicle slid into the deep shit-filled lagoon.

Randy's death shouldn't have happened but it did. His family never got any compensation, even though a union took up the case but it stopped after a while. Randy was buried on the farm near the shit pit and the cows and the broken front loader.

Maria got involved with farmworker movements after this, and now she works to help protect the rights of others on the farms they work on.

I used to be an investigative journalist until a decade ago, and my trip to meet Maria marked my return to this form. When I heard Maria's and Randy's stories, I was so enraged, so charged, the essay I penned wasn't so much an essay as a rant. I was angry at the world for the way it had dealt Maria and Randy such cruel hands. Now that I am calmer, I understand the anger I felt but also understand the situation in a broader light. Maria suffered but she is proud of her life and of what she has built and what she has done. She told me that the American dream is a real thing, though it is harder to achieve than she first thought. She has a dignity about her and strength too. There has been a cost to her gaining this dream in the form of her labour, but she has made a better life in her adopted country, better than the one she would have had in southern Mexico.

Despite the thievery, the sexual abuse, the death of Randy, Maria is a happy person, a jovial person and a woman of morals. She still has a strong sense of what is right and wrong, of the dignity of others, and that to me counts for so much.

I think about her and the other workers as I trundle around the yard. Sometimes, I look at their pictures and wonder what they are doing, all these labourers who I came to know in my weeks on the road.

Before I was a writer, when I could not get a job as a journalist, I worked on building sites in Sydney. The labour was hard, and though I suppose there was some honour in it, I could not see it. It needed to be done so that I had money to keep me going in that expensive city and then to get me back to Ireland so I could write, and so I did it.

I remember the cold early mornings and the hot days I spent on the skyscrapers. I remember the heatstroke I suffered under the hot Australian sun, and I remember the small wages and the employers who didn't pay too. On one job in Redfern, on a huge tower where I worked for many weeks, I would buy my smokes and a Vietnamese banh mi roll at lunch. I remember not telling the other labourers that I had once been a journalist in the city, that I had worked in air-con-filled offices and sipped flat whites. I did not tell them any of this nor that I had the dream of being a writer. I kept those things to myself and instead talked about football or rugby league or anything to pass the day. It was a simpler life but I knew I was on the lowest rung of the building trade on the site. I was unskilled and dispensable. I never refused the money, though.

As a writer, I work each day on my craft. I write on the subject of work, of labour; it fascinates me because for so much of my life I have had to practise it myself. Now on the farm, I give my labour up to these sheep, these lambs, happily. There will be no money in it for many months until the lambs are grown and ready for the market. When I sell them, my labour will garner me a small profit if I am lucky. My own money from my own endeavours. There is something good and honest in that. A book can be like this, too.

Labour, I suppose, is part of life. Maria taught me that lesson. Sometimes we give it freely; sometimes we charge for it. Work is part of our time, and some of it is good and some is bad, but Maria's time, my time, Randy's time: it is all important. It all has value. Writing, farming and building – it doesn't matter what the job is.

I close the picture of Maria and put the phone back in my pocket. I have not finished all my jobs. I gather up the last sheaves of hay in my hands and continue with my work. Maria will never know my sheep shed but I think she would enjoy it.

Labour is part of our gift and part of the curse of being alive. We rejoice when we have enough of it, as it gives us purpose. It is part of the 'sacred rights', as Pope Francis calls them, to land, labour and a home. To have its lack is to know hunger and sadness. Maybe my years as a migrant are coming back to me in a new form because I am happy now to be a shepherd, to offer up my body willingly. It is hard work but it is pleasing work because it is for myself. Not all work is like that. Maria has taught me that. Labour is part of life. Labour *is* life.

We must love
our home

t is the dead of winter and I am outside today with my father, away from the sheep shed. Now that the sap is down and most of nature is dormant, it is the time to plant. I have wanted to plant a number of trees by the nearby fields for some time. We restored this ground years ago, but until today it has been missing a part of itself. It needs trees and bushes. It needs these plants for many reasons: for cover, to act as fence lines, for art and, perhaps most of all for me, as reparation.

I have flown hundreds of thousands of miles in my life, from the Yucatán peninsula in Mexico to the metropolises of Japan. I have contributed my share of carbon dioxide, monoxide and other vapours to the atmosphere. In doing some preliminary sums, I work out that I have travelled over 250,000 kilometres in my lifetime. More than anyone in the history of my family. Why I roamed is a question I have long asked myself. I left Ireland as a young man for a six-month exchange programme – the result was nearly a decade away from Ireland. I did not plan it that way: life simply happened, and I was on board for the adventure of my twenties, from camping out in the Australian outback, hunting buffalo, at one time living in a penthouse in downtown Toronto years later. I travelled because I wanted to be like some Hemingway character experiencing the world in all its highs and lows. I loved it all. I loved the smell of the airport and the feeling of the jet engine powering on to lift us into the sky. Perhaps wanderlust was in my veins. I loved the feeling of being away, of opening myself up to a strange new world. It was freeing and exciting and everything that one's twenties should be. But now I am rooted; I am like the great spinning wheel at last come to rest. After all these years of wandering I find travel hard now. Isn't that a funny thing.

For my reparation, to offset the carbon released, I would need to plant thousands of trees. Reading up on the subject from the Tree Council of Ireland, I learned that a car that drives an average of 18,000 kilometres in a year emits some 2.14 tonnes of carbon dioxide in the process. It would take 178 evergreen trees, in their full maturity, to absorb that. Planes are much the same when broken down to the individual level.

The trees my father and I plant have a dual purpose; they are a reparation and a peace offering to the earth. I plant them to say sorry but also to pay dividends to the great mother. After my ten years as a traveller, it seems the right thing to do.

I ordered the trees from a nursery and they came via a delivery van. It's amazing to know that nature can be so readily available. As I take the spade in my hands to dig up the earth to make the hole for each tree, it occurs to me that every action in this modern world of ours has an impact on the environment. Every car journey, every lit fire, every warmed radiator is releasing carbon into the atmosphere. Of course, to think totally in this way is a form of madness. There are many modern conveniences that have made our lives better. The tractor has changed so much on a farm: where before it would have taken teams of horses and men, it can plough and lift and carry the round bales of silage, which can be made in bad weather and used whenever one wants.

But there are other things that came that were not good. We drained land and bogs to make use of them but now we see the full impact of what we did. It has all led to our *Mu* lifestyle. A Japanese term for lacking, or pure nothingness that I came across while reading recently. Mu puts us at odds with the great mother. When we occupy the nothing space, we do not let nature, the connection space, come into our lives. The more we evolve in our civilization, the harder we become on the earth and so the more Mu grows, until the space between humanity and nature has become a great black hole, on one side the earth denuded and prone while we are on the other, ourselves alone in our concrete isolation. The trees are my best hope of paying my dues.

As a species, we must concede that the trees were here first. We removed them from their land and we cleared their forests because we thought of them only as fuel or raw materials. Once, this country was covered in trees. We were the great forest-dwelling people of Europe. We interacted with the trees and perhaps saw them as gods, as other early people did. For example, in our folklore, there were seven magical trees in Ireland, each with its own properties and powers. Trees were part of our ancient life. However, by the turn of the 1600s, less than 20 per cent of Ireland was covered in forests, and that fell to 1 per cent by the end of the nineteenth century, making us one of the most deforested nations in the world. How did they go from our fathers to our victims? Some blame the conquering English and their need for timber for boat-building, whilst others say a rapidly growing population cleared more land for farming. Whatever the reasons, the great forests of the Celts are long gone. The Irish countryside, so famed and green, is a created landscape. There is no Forest of Dean here; there are none of the giant redwoods of Northern California. Indeed, despite afforestation programmes and commercial planting today, we are still below the European average for numbers of trees.

We have decided to plant birch trees and sycamore. The birch trees are an emotional choice. Before I was born, my uncle John, the second of the Johns in the family after my grandfather, planted birches in the ditches of the home farm. The trees grew quickly and provide shade and comfort now to the livestock. The birch is not a long-lived tree, reaching maturity at some sixty years, so the work of Uncle John will need to be supplemented by the next generation. John is now dead and, leaving no children of his own in the world, the trees are his living legacy. When I look at them, I think of his quiet easy way, his walk, his tobacco pipe, his broad workman's hands. These new young silver birches will grow quickly: I have often seen them in the bog lands where they have quickly colonized empty ground and grown thick and strong. I like their long slender arms and leaves. They have been the tree of my childhood, the tree that gave the farm its name: Birchview. The birch trees have provided many a nest for our birds on the farm, and I have seen crows and blackbirds take up residence in their tops.

With the sycamore, I have a different relationship. Years ago, when we gained the hill farm of Clonfin in the east, its ditches were full of grand old sycamore and beech trees. The land had been part of the Thompson estate, and the farm itself had belonged to the game-keeper of the estate. Though we are separated by the epoch of time, that man planted those trees and they have supported squirrels and birds and many insects.

I struck upon the idea of planting sycamore myself a few months ago, when my father told me about seeing a red squirrel for the first time in the upper ground. It was a sight worth talking about, we had not seen them there before. In the planting of a big tree we must be careful. Like children, they are vulnerable to sickness and wind, to poor soil and, perhaps most important of all, to livestock. Last winter, my father planted a hedge of laurel and the sheep ate every single one! So we must be safe with them.

Planting a tree goes in the opposite direction to our nature because it is not about taking something but about giving something back. So much of our labour on this earth is about taking, it is 'extractive', as the American writer Wendell Berry rightly calls it; indeed, there is, as he says, such a thing as 'extractive agriculture'. That is a thing worth thinking about, how we can be not in balance with nature on our farms, I suppose.

For a great many of us planting a tree is an alien concept, something that is done by a politician at the start of a new housing project or in remembrance of a great event, often historical in nature. When it is an extra fee on a flight ticket for, say, carbon offsetting, the planting is done away from us, out of sight, in the nameless place of somewhere else or a sort of now-restored wilderness, though one never sees these places. Indeed, many now say that carbon offsetting is a form of greenwashing and that the act distracts from companies actually doing real work to offset their emissions.

A close friend has taken matters into his own hands and has refused outright to fly. If he must travel now, he does so only within Ireland by car or wherever he can reach by boat. I admire his intentions. He wants to do good by the earth and in the doing has struck his bargain with the great creator. If we are lucky enough to be able to fly, we must consider the damage we are doing.

For so many, being around plants is not the norm. Sometimes I think of the big urban jungles and how nature can at times only be a small tree or a flower on a windowsill. Indeed, residents of some cities, such as parts of LA where I have been, are not only cut off from plants: they are cut off from fresh produce in their stores. They are 'doubly impoverished', as John O'Donohue coined it. Concrete is great but it cannot beat an earthen path.

Being with trees, being with nature, as I am today makes me feel happy. It reminds me that there are things on this earth worthy of our attention. The trees let me know that I am alive and that money, travel, our modern way of life might need to be rethought. I cannot escape my own argument here because I too am guilty of these modern trappings. I too forget about nature but I am trying to remember. I am trying to recall.

This all raises the question: is farming part of the problem? Agriculture places a big demand on resources around the world. We cleared the indigenous worlds of North and South America and Australasia so we could put our livestock on it. Wherever the European went the cow and the sheep were used as an excuse for our encroachment: the animals needed the land, not us. It was always a flimsy argument but the sheep booms of, say, Patagonia in the late 1800s point to the European ideal of going big or going home.

Patagonia is a huge territory of some 900,000 square kilometres. It is a tough land, cold and harsh in parts, well suited to a woolly animal like the sheep. The guanaco, a llama-type creature, had survived there for millennia. The lower south of Patagonia was composed of natural grassland so it was ripe for commercial livestock grazing, and in many areas no deforestation had taken place to work the land.

The first farmers came to Patagonia from the Falkland Islands in the late 1860s with their sheep, according to the writer Nicholas Shakespeare. The sheep that arrived on Tierra del Fuego and Santa Cruz came along with Scottish settlers. As John F. Bergman of the University of Alberta wrote, the Scots were 'expert sheep men and accustomed to a rigorous climate'.

The introduction of sheep had a huge effect on both the country, the inhabitants and the environment. The first sheep farmers worked the animal for wool, but soon meat was also on the menu due to improved transport and refrigeration methods. The sheep farms, also known as *estancia*, spread throughout the region as land concessions drew Europeans south. Amongst them were Welsh, English, Scots and Irish people who had left one cold land to inhabit another. Indeed, a large contingent of my fellow men and women of Longford emigrated to Argentina at the time, and the links with the nation still hold today.

Great sheep drives were organized to get animals, people and supplies to those southern land concessions. By the 1950s, according to Scottish writer Karen

Emslie, the sheep population of Patagonia numbered some twenty million.

The impact of so many sheep on the landscape resulted in what the *Guardian* newspaper described as 'amongst the most damaged grasslands in the world where some twenty million acres have been abandoned'. In our quest to produce ever more wool and meat, we have pushed mother nature. Pushed and pushed and not thought of tomorrow.

When I think of sheep I think of that loyal and lovable animal, that member of the family of man, but I also think of the environmental damage they have done. Patagonia wasn't the only land to be harmed by sheep. In Australia, my old home, the vast mobs of sheep – for there are no other words for the sheer numbers of animals – quickly grazed and degraded the native grasslands from the Riverina in New South Wales all the way to Adelaide. By way of example, in Tasmania in 1827, there were 436,256 sheep on the island. By 1836, that population had risen to 911,357 – twenty sheep for every white person in the colony according to historian Robert Hughes. The indentation of sheep's teeth allowed them to eat lower and deeper into the grass and then their hooves compacted soils, further denuding the fragile grasslands. The Australian soil had never known hoofed animals before the arrival of European stock, and soon cattle, sheep and rabbits had destroyed great swathes of prime grazing grounds. Today, these native grasslands are under extreme threat with only 1 per cent of south-eastern lowland native grasslands in decent condition.

Grazing is an important part of Australian agriculture, and some 336 million hectares, or over 40 per cent of the total land area of Australia, is devoted to grazing land for livestock. The logging of old-growth forests for timber, and in some cases for more grazing land, saw huge sections of forest cleared in the southern Australian state of Victoria, where only a small percentage of true old-growth forest remains and even that is under threat. As Don Watson, writer and speechwriter for former Australian prime minister Paul Keating, said in a recent lecture: 'You can't destroy a forest without there being regret.'

There are now various attempts underway to restore the two different landscapes of Patagonia and Australia, but sheep farming has left its impact on the land to this day. That is what people can do when we take nature into our own hands. What we must face up to. But we do not need to go to the ends of the earth to see our impact. The sheep population of Ireland was once at 8.9 million and they have degraded grounds and soils.

Today, the numbers are lower as we try to come to terms with our impact on the earth.

I wonder if the sheep knew of all we have claimed for them would they be happy with us? They are not a greedy animal. It makes one think, indeed, what all our animals would think or say of us if we could really communicate with them.

A few weeks ago a First Nations friend and artist from Canada, Adrian Stimson, sent me a care package, including T-shirts from his tribal nation, the Siksika band of the Blackfoot from the state of Alberta. Included in the package was an essay about the Siksika by Ben Miller. He recounted how the Siksika have a unique way of bonding and tying themselves to the land through their dances and initiations. The *ako katssinn* or sun dances marked the initiation of young men, where dances and body piercing took place. The *ako katssinn* rituals and many others of the tribe are referred to in the essay as memory capsules that Native people use to encode their belonging to the land. The memory capsules, as the Siksika said, are a way to reconnect them to their roots.

I think that through our own actions on the land we are encoding ourselves within the life of the farm. The trees are a sort of memory capsule for us. Being around trees, as Da and I are today, makes me think now to remember when they were planted, what we were up to and who was there. Thinking of trees also makes me think of times past: that then we lived a life that was more in tune with nature. Going back is not always a bad thing. There are methods in that distant land that made sense, methods that were good: farming within the seasons, waiting for the bounty of nature and knowing that we could only have our tomatoes and fruit at a certain time of the year. In Australia, when I lived there, the nectarines and cherries only came once a year, when they had grown and matured naturally. It made me appreciate eating within the seasons. It was in line with the natural world. We must understand, too, that we can only 'put so much weight upon the humble beast's back', a phrase my father said to me one day and that has stayed with me.

This is why, I suppose, my father and mother have taken the decision to convert the farm to organics. It has made me, their son, very happy. In organics, we are breaking away from the greater system of agriculture. In organics, as in the planting of these trees today, we can go easier on the planet. Already organic shops and markets have sprung up around the area, from an organic burger restaurant to fresh-food delivery boxes of fruit and veg or, in the case of another friend, meat boxes from his organic herd in Sligo. These businesses do well and it points to the fact that society, too, is interested in organic farming. It may come at a premium price but there are many who see the value in it. My father tells me that if people could just spend an extra few euro each week on organic produce, they would be keeping many farming families alive.

In this way, the people of the cities support the people of the land. We are in a communion of commerce together. It brings me great joy. It gives me ideas about what we can do on our farm, what produce we could provide. There are the cows and, of course, the sheep but sometimes I think that the written word and the farm can combine – that perhaps people might want to stay on the farm and work creatively. This could be our farm business all of our making, a new fruit of the land and the word.

Reading the farming supplement of the paper recently, I learned of stock numbers declining on farms in Western

Canada due to drought. One woman had to sell off fifty-one head of her cows, including twenty heifers, which were future breeders on her farm. The column said that she was selling her future off. That was a line that struck me deeply. Without a future, what are we left with? Not a blank canvas but a bereft canvas, one absent of all hope.

The western states of North America are living through drought at the moment, and friends there have experienced wildfires, which, they tell me, are due to climate change. Indeed, so bad were the fires last summer that an entire town in British Columbia in Canada burned. South of the Canadian border, a third of US cattle are in drought, and culls have happened as feed has run scarce. In some southern states of the US, like California, it did not rain for months, the same article noted. Indeed, they experienced the worst drought in twelve hundred years in the south-west. They call it a megadrought. That is a scary term, a life-changing term.

On a trip in California in 2020, I was amazed to find alfalfa grass growing in the desert, using clever irrigation. But at the same time, the river systems in California are heavily polluted from fertilizer and agricultural run-off, as has been well noted. It is a scary thought when one imagines the future in these places. The problem with these modern farms in the settler societies of the world, from Brazil to Australia, as that newspaper article detailed, was that people did not respect the fact that the climate was changing. Of course, many of these farmers came from Europe originally, and they imposed their system on a landscape that had been worked in a fundamentally different way for thousands of years by its original inhabitants. No thought was given to them then and none is now. The irrigation networks of, say, the Hohokam people of Arizona are only a footnote. The villages of Aboriginal Australians were wiped off the map by hungry graziers, as Australian historian Bruce Pascoe has detailed in his book *Dark Emu*.

In Ireland, our winters are becoming wetter and our summers drier. In recent years we have experienced what can only be described as rain bombs, which have brought increased flooding to vast sections of the country. In 2016, I remember hearing stories of farmers who had to sail small boats out across fields to feed stranded cattle. Droughts, something we are not used to, are starting to occur. In 2018, a heatwave raged across the UK and Ireland and led to crop failures and water shortages. Living through the summer in Europe in 2022 the Iberian Peninsula had its worst drought in twelve hundred years (the same as the previously mentioned megadrought in California). Rivers in Germany usually used for freight faced severely reduced water volumes. Recent studies in Ireland point to some of our waterways losing 30 per cent of their volume due to climate change. I worry about what the future holds for us all. I worry more for the next generation, who will inherit an earth that is suffering under the weight of our overuse.

By turning organic we can, I hope, farm and work more in unison with the world we have been given. It is not some strange idea, rather it is slower living. If farm families can survive in a rural landscape like our own then more families can survive in a rural economy, and the money generated in that economy stays in the ecosystem of that area. Rural communities need family farms; they need a living to be made from the soil, an equitable living, something that is increasingly getting harder to achieve. To put it in perspective, in Ireland in 2021, the average income of a sheep farm was a little over €20,000, and this was an increase of 16 per cent on 2020. Imagine all that work for so little return. There are the costs of input – land payments, rents, medicines, vet bills and so on – and the fact that lambs sometimes fetch low prices. When I sell these lambs that I am bringing into this world, I will not be a wealthy man. But people continue on, they persist, because they love this way of life and the bond they have with their communities and their animals. Our actions now prevent rural decay in the future.

I have seen the end result of dying rural communities, from America to southern Europe. I have seen what rural decay can do. Travelling through the middle of Spain some years ago, where rural depopulation is a major problem, I saw a possible future for our home. The young were gone and only the old remained. The young had been educated to leave. They are never educated to stay. In their departure, or absence, was left a community such as this small town in which I was to spend the night. I can no longer remember its name, but as I walked into town, even the fields in that land were empty, bare of trees and shrubs. My journey took me along an old road. The town bore the hallmarks of having once known wealth, having once known industry. Even its churches, some of the most beloved buildings of the Spanish, were decaying, the masonry giving way to time. The question of the rootedness of the churches is akin to the rootlessness of the young. They go to the regional centres, or even further afield to the capitals of southern Europe and London, while the places where they once prayed fall to the ground.

As I load another tree into the ground, I think that perhaps it's all to do with foundations, with roots. If we do not have a hold on the earth, then what do we become? Ireland, quiet rural Ireland, is becoming like this in some ways but it is not so bad, not just yet. There are still young people here but our villages in some regions are dying. It happened slowly: first they took the post office, then a garda barracks, then the rural bank, and now even the pubs are teetering on the edges of extinction in many places. The question is not are they dead now. No, it is what will they be like in twenty years, when the members of the last generation are gone?

The media have talked about it for long enough that the people are waking up to it too. In my hometown, I have seen new people come and set up new businesses. High rents in cities were one cause and the search for a quieter way of life another. Migrant communities have been established and now we have Eastern European- and African-owned businesses in our town. It is a good thing, a bright thing. I would hope that can happen in other rural communities. Once I wrote an essay for the radio on things I missed from the city. Top of the list was sushi, but recently a sushi restaurant opened in our local rural town. All things come to those who wait!

In that small town in Spain where I spent that night, where the stonework was faded, where the road was potholed, I thought of my rallying cry: that it is our job, that of the people of my generation, to turn back the clock (perhaps job is the wrong word because, in many ways, it is a calling). We must call the young to return, to save both the present and future kingdoms. The future of these places, in America or Spain or Australia, is an interior empty of her people. Who is to farm the land? Companies with employees? Bill Gates is now one of the biggest owners of farmland in the US, so maybe. For us on Birchview farm, we have made a decision to stay on the land, to work it in a new way and hopefully, by the doing, to be able to live on for another generation. We must not be greedy with the earth or the animals. To have sufficient funds to keep going is all one asks. It is enough.

And what of the rural communities? What can we do to ensure their survival? Recently, a group of friends and I set up a book festival in Granard, County Longford. The festival brought writers from all over the world into the north of the county for a weekend to share their ideas with the rural community. We called Granard a booktown, after the international model, and got help from a sister rural town in Scotland called Wigtown. Our first weekend sold out, and crowds came from all over the island, not just the environs of the county. Money was spent in the town that weekend and proved that rural communities are open to new ideas just as much as big urban centres. Books have been our means to entice people into our beautiful home and, so far, it has seemed to work. Already there are plans for the next year, and the writers will come from the ends of the earth to see what we are doing for our future.

I fill the soil around a now-planted tree with my shovel and, judging it to be right, myself and Da move on to the next patch. There are now plenty of trees planted and I am feeling the benefit of my work. If we are to prevent rural decay, we must learn about growth like the young trees; we must learn from the cities and bring back the power of their ideas. There are people in rural communities who like theatre and fashion and coffee just as much as those on any high street in any city; their wishes are the same as any urban person's. The pandemic has disrupted the pull of cities: the rural has had a reprieve. A rethink is possible because, for the first time since anyone can remember, our youth have returned from the likes of Dublin and London and New

York, even my beloved Sydney. The global family can come back and with them a new future could be born.

In a way, I think, the trees I am planting are a means to reconnect, a way to rebuild. When I am old, the trees will be maturing nicely; when my children or nephews and nieces come to the land, they can remark on them as I once spoke about my uncle John's work. Our hands are only upon the earth for a short time. We must invest that touch wisely. We must live more intimately, so that we can dream big.

Chief Joseph of the Nez Perce tribe, a man Father Sean told me about many years ago, said, 'the earth and myself are of one mind'. I am coming to that great man's understanding: that the measure of the land and the measure of our bodies is the same. That, as he said, 'good words do not last long unless they amount to something'. We are called to act now, not tomorrow.

I pick up the next tree and Da digs the hole. There's something right in this work, something whole. It is my reparation and my redemption.

Death is
part of life

This morning we had a lamb die. He was born wrong or maybe it was that I was not right this day. The birth was laboured; the ewe had been at it a long time – I could tell from her warm passage. It was hot to the touch: a sign that things were not good. When I took the tangle of legs and set the lamb to rightness, I brought him into the world. This is always a moment of nerves because I do not know if life will come into the lamb's form. I say 'come into' because that is what it feels like. That life comes into the small form from somewhere else, somewhere eternal. I know that life came with this lamb from the world of the womb to the world of the present, but I have seen so many struggle this way that it makes me think otherwise. You must always think about that life energy in a lambing season.

Life, death, they are real things. Both are big and strong. Both are stations in our time on this planet: a beginning and an ending. It is the same track everywhere. We cannot avoid them in our soul's journey on this earth. We are called in conception but we are also called to die. Both are calls to another world, another mode. When I pulled the lamb into the world this morning, I had a rope of twine around his head to help bring him forward. The ewe's cervix was tight and I did not know how I would get him out. That was a battle that I had not counted on but it was a battle that I knew from experience. To keep one's head cool is the right measure in these moments. To follow the rules of birthing are the only things that can be done. When I took the lamb out after twenty or more minutes of work, his eyes were closed and the yellow birthing fluid covered his fleece. His nostrils did not move and his head did not shake, and so I took him and swung him back and forth gently to help bring the fluid from his lungs. I laid him on the fresh bedding, cleared his nostrils and put a piece of straw up his nose – this sometimes clears the airway and helps the breath to come in. But after the briefest moment of life, where his eyes opened, he then seemed to fall back into the world of the dead. The shock of life, the shock of the new world, was not coming into him. Perhaps I had not been quick enough; perhaps he was never meant to live. I do not know. There was no life there.

I rubbed his small lungs and then shook him gently again, but he did not come alive as I have seen so many times before. He lay motionless and then, after five or ten minutes, I understood that death had taken him, or won this battle, or whatever term one wanted to call it. The ewe turned to lick her lamb but she could not bring him back or forward.

'There is no manifesto but the breath.' The phrase comes to me from a long time ago, from a poet I once knew. It was he who first got me to perform, he who first showed me that the spoken word was a powerful thing. A year or two ago I woke in a dream of the man. I had not thought about Raven for nearly a decade or so. But this summer night I woke and he had been in my dreams so loudly, so fervently, that I took to my phone in the middle of the night and looked him up.

I found that he had died that day, that he had drowned in the waters of a lake in Eastern Europe while on holiday. Thinking this so strange, I woke my wife to tell her. She had met Raven once, but it had been so long ago I was not sure she remembered. It was dark in our cottage; the news hung in the pitch blackness; there was no light anywhere in the world in that moment it seemed. The light of the world had been snuffed out, quenched for me, in that instant, and I knew that he existed now only as a memory in the hollow of my mind. My wife fell back asleep but I could not, and so in the dark room I thought deep currents of memory. In that imaginative space I thought back to my college days in Dublin, where I'd attended a spoken-word open-mic night each week in the city's arts district of Temple Bar. It was the early days of the beat scene in Ireland and it was all gathered around Raven, a black American who could rap and recite, who could spit a cypher and who was the elder statesmen to us all.

I had never thought about poetry as an art form for myself before meeting Raven. My world had been consumed with books and music and, to a smaller extent, my studies in journalism, which had not as yet stoked the flame of an investigative journalist in me.

Thinking of Raven now and his beat poems, the words have nearly all evaporated. I remember only some lines. Lines about the fight to be heard. I learned about politics, in particular black politics, from Raven. On quiet days when no classes were scheduled in college, I would call to his house and he would educate me on African American issues and history. In his little studio at the rear of his house he had a pyramid of pictures of black leaders. He told me about books by these great authors and thinkers. They had all been unknown to me before, and it enticed me to read more widely. My politics, such as they are and were, began in his old Georgian house. I read Malcolm

X first and found the radicalism of his words charged me in the way of the dispossessed. Malcolm X's journey towards Islam was a reckoning ultimately with the self. I was attracted to his spiritual journey. I was attracted to bravery. From this point I came to understand the struggle of the under-served, the forgotten and the maligned. From Tibet to Native America, it awoke in me an awareness of the need for people to be heard.

It was at this time that I, too, began to think of my own name and whether it should be written and said in Irish (something I ultimately did try out years later), going by the name John Sean and later Sean Ó Connail. The name change was an act of politics. It was also an act of reclamation.

After reading all those books by those great thought leaders, I found myself a different man. Perhaps the story that stays with me the longest from that time was that of Alex Haley, the writer of *Roots*, a novel that embodied his search for the history of his family. *Roots* was a doorstopper of a book that I brought around with me for many months, dipping into it to find new routes to wisdom. It was a book that travelled with me to Australia and one that I dipped into again as I learned about Aboriginal Australian rights.

Black nationalism was something that Raven and I talked about too. How groups like the Black Panthers or the Pan-African leader Marcus Garvey fought for a new future for the minority and how they had produced change despite it all, in America and beyond. Resistance was important, resistance to entrenched views, resistance to a dominant culture that served to hold others down. Change did not always come so easily, Raven said. It was like poetry, he cautioned: it wasn't a poem until it was printed; it wasn't poetry until it was read. It wasn't politics until it was enacted. Raven showed me a way into Irish nationalism. In a way, the stories of the struggle for freedom of another people opened me up to the struggle of my own. It made me realize that it was a long and storied tale and that it, too, was part of the story of the world. We were part of the wounded of history. I read about Northern Ireland, about the civil rights marches. I came to understand the fate of my people, something which I had hitherto not considered. Our history had not ended with the War of Independence. It was an ongoing journey, a story still writing itself.

Raven died in August 2019. When I had my dream. When I checked on social media, and found out that he had died that day. It had been the breath, the fight for breath, that had taken his life. When I think of the death of George Floyd I think, too, of that notion of the fight for breath. How nine minutes and twenty-nine seconds could change the world in the first summer of the pandemic. That the politics Raven taught me had not ended, that the struggle of African Americans was not over, that the fight was continually present and would take a long time to come to its conclusion.

I last saw Raven years ago. I had returned from Australia to Dublin for a brief holiday of three days. Walking down Dame Street in the city centre, on a dark evening, Vivian, my then girlfriend, by my side, we'd bumped into him. He looked older. When I think of it, he was probably not much older than I am now.

We talked intensely for ten or twenty minutes and then Raven said that he had to go. He gave me his phone number on a piece of paper and I promised that I would call him before I left. The trip was one full of meetings and new projects and before too long it was time to fly to London for a television production. I never did call him and we never spoke again.

I had not thought of Raven for many years but I had never forgotten his statement that it was not poetry until it was read. It was something that I would mould and amend to tell students and younger writers: that the book was never finished nor the newspaper article written until it was with the reader. That it was the reader who completed the circle. I remember the words now as they call out beyond time and my hurt.

In the dim light of the shed I see now that the breath will not come into the lamb. Life is everywhere around me but not in this lamb. I cannot will that force into him. The charge of electrical life is absent. I check the ewe's udder and both her tits are working and milk is coming forth. We will get a foster lamb for her from someone local or perhaps on Done Deal, the buyers' and sellers' website. We cannot have a ewe without a lamb; we want to have something to show for the season. It will not be her lamb but we will wrap the foster lamb in the cleanings and fluid of the ewe to trick her into thinking that he is. It will have to be quick work because this sort of replica of life can only be done on the day of the birth. I have seen it tried later than that and the ewe will not take with a lamb and must be put in a foster crate (a metal contraption to hold the ewe in place by means of a headlock), and even then, there can be protestations.

I take the old lamb and place him to one side. We will dispose of him properly later with all the right measures. As I carry his lifeless form, I think for a brief moment of all that I would have said to Raven had we met again. It is not the death itself; it is the absence, the never-seeing-again nature of the departure from this life that hits me the hardest. It is a total incompleteness, a fracture that just stays broken.

Death is part of life. That is my lesson today. I have learned it many times in many ways.

All mothers are a link to the great mother

n the mornings, Mam is up early. She walks out to the yard to feed her hens and then looks at the cattle and sheep to see if any new life has come in the dark of the night. Each morning, before I am up, she texts me to say 'happy day'. The text comes whether it is a bad-weather day or not. Some mornings she has news of new life – a calf, a lamb, even a foal – and in recent months she has learned how to take pictures on her phone so I see the new babes in all their glory.

By the time I get to the farm, she is already hours deep into her day teaching at her Montessori school, and we may meet for a quick chat at noon or a coffee along the way. I tell her about the sheep, of the happenings in the shed, and she delights in the news.

Margaret, or Mam as I call her, was born on a farm a few miles from Birchview, in the townland of Aghakilmore in north Longford. There were five children in the family. They say the people of north

Longford are industrious workers: that the land itself made them good farmers. There, in those fields, they learned to mould and shape the earth. They have shaped it for hundreds of years. I think those fields were the first great works of beauty of the people long ago; they cleared the ground to make their dreams possible. The fields were everything all at once. I feel myself akin to those north Longford farmers: they are my ancestors.

My mother has taught me many lessons in life and, of course, our parents are our first real teachers, but I have learned deeply from her many times as both boy and man.

My mother's first memory is of the byre (cowshed) of the farm in Aghakilmore, where her father Thomas once gave her a penny. Through our chats over the years, that penny has taken on great significance. It was a token of love, but also a marker, a sign that she was doing well.

Granda Reilly was an industrious man, a man who, even though I never knew him, seems bound to me now in memory and imagination as the grandfather I would always have loved to have had. Thomas was an entrepreneur – a vegetable man, as my mother puts it. His land all around the house was planted in rows of all kinds of vegetables, from cabbages to kale. He was a busy man and employed two local workmen to till the soil with him. There were animals on the farm, but as my mother says, 'Daddy didn't know how to milk a cow – his love was his plants.'

He worked hard at his job and would often be up at four in the morning to go the markets and fairs with his plants for sale – so early that the patrons of the pubs and hotels would still be enjoying their night out. He would hear the sounds of show bands still playing as he made his way to Ballyjamesduff or Longford town. He would return with green pound notes in every pocket and spill them out into a basin, where they would be counted and saved. When he gave my mother that penny, it counted for a lot. In a time when no one said 'I love you', the penny said and was everything.

Granda was also a great man of learning and in the evenings he read books of all kinds, from history to westerns. Father Brennan, a local priest, used to bring him new releases. My mother says that my love of words first came from Thomas, that a love of reading was in the genes of the family from then.

Annie, his wife, was a graceful woman who had come from a farming family too. She had been a priest's housekeeper before marrying Thomas. Indeed, many of her sisters had because it was a good job back then. She had come from a houseful of five girls and three boys, one of whom become a priest. Her marriage to my grandfather was an arranged one, by a matchmaker, as was common in rural Ireland in those days but, according to my mother, it had been a successful one and a good match.

Annie took care of the livestock and the children, who all came in quick succession as their family life took root in Aghakilmore. Life was good, and though everything had to be earned, as my mother says, they didn't want for anything. On days when her father went to the nearby town of Granard, he would return with chocolate bars for the children, each of them getting a different type. My mother would receive a Kit Kat.

The accident happened out of the blue. Granda was kicked in the stomach by a calf. The kick, or maybe it was something else, we will never know, turned into a pain and the pain turned into stomach cancer. Thomas was in his mid-forties.

My mother was only a child when her father underwent surgery for his stomach, but though the surgery was a success, the pain, the cancer, continued. As my mother remembers it now, he was sick for a few years as that terrible disease began to strip him of his power and his strength. When the threshing machine was in the townland to work in the summer, Granda had not the power to till the fields nor make the hay and other men carried out the work for him. Granda was a man of action and had his own pride and he did not like to be unable to work. There was, as my mother remembers, an anger in him, an anger that he had gotten sick, an anger that he was weakened, an anger that life was moving beyond or past him.

I do not know whose idea it was, perhaps no one alive now knows for it happened so long ago, but the neighbours banded together in a *meitheal* to raise money to send him to Lourdes, the place where the Virgin Mary appeared to Bernadette so long ago and that has since become a place of Catholic pilgrimage. It was hoped that, as a place of miracles, Lourdes would cure Granda. People's faith ran deep in those days, deeper in ways than now. There in that land, it was thought, God could meet us in the good and the bad.

A nun accompanied Thomas to France, and when he returned, my mother says that he was a changed man and the anger was gone. The miracle, as she sees it now, was not in health, but in acceptance. When they went to see him in the bed near the end, when he was fading away, he was at peace.

He died aged fifty. Gone were the chocolate bars and the other gifts and the single big pennies by the byre. My mother was seven years old, seven cabbage-plant seasons of the earth. Often in life there are two sides to our journey: the before and the after. That was the way for my mother when suddenly her mother was given the task of raising five children and running a farm. Granny was not a horticulturalist and so the cabbages and kale were put aside and a new business started. Before long, my grandmother had twenty-five dairy cows, a big undertaking in those days. She farmed the land and rented more and, together with the children, she filled the milk canisters for the creamery every week and got the milk cheque every month.

Life was made harder for my grandmother because she was a woman in a man's world, but she held onto her land and to her children, something that in the Ireland of that time was an achievement in itself. My mother remembers stories of other single mothers whose children were taken from them and their land divided out.

It was a time of war against women in Ireland.

My mother remarks that her mother gave her the appreciation for a job well done. That is something that has stuck with her through her sixty years on the planet, through the creation of her Montessori school and the farm. Though life gave my grandmother a hard road, my own mother sees the greatness in her. It has taken her time to see that, for whatever our age we are all still children in our parents' eyes.

When my mother walks out to see the animals on her own farm each morning, she enjoys it – even the mucky wintery days, she says, for she knows the animals are safe and warm in the sheds. She knows sheep and cows; she understands their ways. For her, it's not a money thing, though money is needed, rather that the animals give her a sense of safety, of rootedness. When she is on the land in the summer and the cattle and sheep are out and their bellies full of grass, she is happiest, for she can feel the connectedness of the animals and that gives her a sense of calm. This calm, she says, comes from her own childhood on the farm but also maybe from something spiritual, something religious. She is a religious woman and it is from her that I received the gift of faith. It started with us saying a decade of the rosary in the evenings as a family. It grew from there. Even as a child, it was commented that I could have a vocation, that I was made for the cloth due to my quiet nature and my love of the environment. They added that there had not been a priest in the family since my mother's uncle, so it seemed time. Though my mother never pushed it, we both found a solace in the comfort of faith. When a cow calved, we took down our statue of Saint Francis, and when we got into a new farming venture, we blessed the animals with holy water. When I went to Australia for the first time, she presented me with a set of rosary beads to keep me safe. All of it made an impression upon me.

My parents did not question their faith, but as I dove deeper into my seeker's quest, I began to ask questions of spirituality. My own journey would take so many turns before I could call it my own; it would be formed by my natural love of the world but also something Celtic, something of the earth. But my spirituality started in that little kitchen praying with my family. 'The family that prays together stays together' as the saying goes. In many ways we have found that to be true.

My mother said something to me once that has always stuck with me: 'When you care for the animals, they will take care of you.' You can fuse with the creature, as she says. She does not believe in the mass production of animals: she believes they deserve to have a life, too. I think she likes my only having the twelve sheep. It is not a greedy number and I can have a relationship with each of the mothers. We both know that the lambs are destined for the food chain, but we understand that we are giving them a good life while they are here. It's economics but also soulful work.

There is a good energy with her on the farm, a female energy, and it is right for I am surrounded by females in these sheep and cows. They are all mothers. They are Gaia embodied. When I am in the sheep shed in the quiet of the day, with the ewes and the lambs, I can feel that fusion she talks of. I can feel that connection. My mother has taught me how to care for the animals, that less is more in this venture and that our relationship runs deep. We, the animals and us, are on a walk of creation together. There is a peace in that. To borrow a line from Seamus Heaney, there is 'a poetry of the living present' in her work upon the land. Each day is a 'happy day' as her texts rightly say. With her, each day is a teaching of the maternal earth.

We must keep beauty in our minds

'All art is a bringing to birth. It is not a matter of creating out of nothing but of liberating what is already there, in the strict sense a labour.'

Pauline Matarasso, *Clothed in Language*

Today I'm spray-painting numbers and markings on the lambs' sides to match them to their mothers. They are nearly all born now and our road is coming to its natural end. The markings have my hand upon them and are abstract sorts of things; they are different to my father's ones, which bear his designs. With the twist of a hand we create our own personal Pollocks and it's got me thinking this morning.

Art is an important thing to me. It is a lesson I learn and live with every day. We have a painting of Saint Francis of Assisi on a simple wooden board in the shed, all the way from Santa Fe, and it makes me happy when I look at it. When I wrote my first book in the Portakabin my brother set up in the farmyard, I had a painted reproduction of Constable's *Hay Wain* in front of me and looked at it as I listened to the real-life cows and sheep low and call from the sheds nearby.

When I was in secondary school, I had the choice to take up chemistry, in which I had some modest talent, or art. It was not a school where I could take both options. I could not combine the chemical and the colourful. I chose art because I was in love with the image. That decision, all those years ago, started a relationship that has stayed with me all of my life. The company of a good painting is as beautiful to me as the chemical equations that go into, say, photosynthesis (a formula I still remember).

Having beauty as part of our lives is an important thing. It is a 'calling' as John O'Donohue rightly said.

Art is not out of place on a farm: rather, it is part of it. We don't just mark the lambs with spray paint; we adorned their mothers, too, at the time of their pregnancy scanning to tell us if they had single or twin lambs inside them. We put x's and circles and lozenges upon their bodies in codes which are themselves artful. Taking a lamb in my hand, now, I take the can and spray him with a blue colour. I have chosen this colour over the red for no reason other than I like it this morning. This act is not done for the sheep, because they know each other by smell, no, it is done for us, the seers.

These marks, they let us know that we have played a part in the making of these animals. Seeing is believing.

As a teenager, I spent a great deal of time drawing images of Celtic heroes and warrior queens. In the summer holidays, I would take pictures of Clonfin, our hill farm then, and try and do watercolours of it, but my artwork has dwindled now. I have no great paintings in me any more. These marks I make upon the sheep give me a sort of creative release. An opportunity to show some flair. The lambs wriggle as I hold them and put the marks upon their small frames but I do not mind.

I have heard from my wife about an artist at work here in Ireland who uses sheep markings from the mountain men of the west as the basis for his artworks. These signs are created so that farmers know their own flocks on the hills and high places. I'm told the artist's work sells well so there may be something in this marking business.

In Ireland, no one is far from a farm; it is hidden in our earthly heritage. The soil of the land is only a generation away from even the most urban of people. We are, I think still, a people of the land. In modern Ireland we are a post-agrarian society, and though we have modernized and industrialized, we still hold a fondness for our farming past; perhaps that is why we return again and again to the theme of farming in our art. Recently, a good friend directed a feature film with his partner about a farming family, which he shot in the next town over. *Lakelands* explored the themes of love and family and community. It was a sporting drama but also a farming drama. It has resonated with so many, its themes having a near-universal quality. Call it a hangover or a connection, but the farm stays with us wherever we go in this land.

Farmers can be, or perhaps always have been, artists, I think. It goes beyond mere markings on animals for it takes creativity to show a beast at a livestock event; it requires flair to train a horse for a dressage competition or to teach a ram to walk around a show ring whilst showing off his well-built frame. All of it is done using our talents. When we farmers see animals, we see more than just the flesh or the price. We can see bloodlines and greatness. Many remember the story of great rams or indeed great dames. The farmer's look is one that is able to peer through DNA and see a living history right back to the first animals that came to this land.

When I look at these sheep of mine, I see the magic of my dreams. I see that the animals have been imprinted with a part of me. It gives me a quiet pride. It connects me ever more to the land, to them and to the ongoing walk humanity has been on with these creatures for over ten thousand years. There is art in the caring.

Art has a real power. We can see animals and so much more.

A while ago now, during the six years I lived in Australia, I regularly made a trip from Sydney to Melbourne. It was my custom to visit my friend Fergus in Fitzroy, but on every visit to that great city I also always took time to do two things: visit Ned Kelly's bullet-proof suit of armour in the state library in the city centre and then stop off at the National Gallery of Victoria, some twenty minutes' walk away, to see Melbourne's favourite painting: *Anguish* by August Friedrich Schenck. The

Anguish

Melbournian's love this painting and have voted it their favourite of the gallery's seventy-five-thousand images on more than two occasions.

It was painted in 1878 by the German/Danish painter and depicts a ewe and her dead lamb. It is a picture of the highest drama, showcasing the inevitability of death, or life, or maybe both. The mother, a good strong ewe who looks like a Charolais breed, not unlike some of our own in the commercial flock, stands over the lamb while the crows are gathering, waiting to feast. She can only hold out for so long before she will grow tired, before nature will take its gruesome toll. Anguish, loss, these are big themes, big events that cannot be so easily put aside.

A few seasons back we had our own moment like this, when foxes took lambs from us. I think of it now as I look at this year's crop of newborns. The lambs were still young at that time and the foxes hungry. They took two, maybe three, lambs: we could never truly be sure of the number because we never found the bodies. I still remember the evening we released them and the big dog fox walked through the paddock calmly and quietly, unhurried by my presence, bobbing along like a little coyote. He had not taken the lambs yet, but we worried about what might happen with such predators present. I am not a man of violence but if I'd had a gun in that moment, I would have fired, for I knew all the work it had taken to get those lambs to that field.

My father called the local gun club that evening who came later and dispatched a mob of foxes. We did not like to see it done, but then it was the lambs or it was the foxes. It was, as my father said, a war we had not asked to start. Still, in the life of a farm there are battles with nature, battles with life and death. We had to protect what was ours; we had to finish the war we had not asked for.

The work of the grey crow is different. Like those in Schenck's image, they are the great pack animal. In my experience, they focus on the young and the weak. I have heard stories from other men of a newborn lambed in the field and they pecked the eyes out of its head while it was still alive and, when it was good and dead, began to eat it, pecking away slowly at its still warm flesh before eating its tongue.

Thinking back now on all those times when I went to see this image of the ewe and her dead lamb, I never thought that someone would live the painting. Art and life can be connected. I am sure that farmer felt as helpless as that painted ewe did when he came across the grey crows feeding on that helpless, eyeless lamb. The grey crows are smart animals. They lived to scheme another day while the farmer was left with the dead lamb, the loss of a season's work. I have read of their work against ewe's too. They are a crafty animal. A taker of life.

For me, the painting is an image that speaks volumes. There is the grief and sadness: an almost Pietà-like moment. There is no sign of hope, no mention that perhaps next year there will be another lamb. It is a blow that reminds us of all the other blows we have had in life. We sit in the moment with the image and are made to feel. It is sad but it is real and that counts for something in an at times unreal world.

Art in the home is an intimate thing, I think. It tells us so much about our personalities. Earlier this season, after buying our house near the farm, my wife and I began to fill its walls with images we cared about. There were maps of the county mounted and framed, so we could better know where we were, a large A1 print of one our Limousin cows, who has now gone to God, a screen-print from an Indigenous Australian artist gifted to us as a wedding present and an image I chanced across online a while back and printed out.

Erich Hartmann is a photographer I didn't know until recently, but his capturing of Dublin in 1964 paints a picture of what he called Joyce's Dublin, for he was trying to capture his hero's city. It's Hartmann's image *Shepherd and Sheep* that I love most of all, a black-and-white photo of a boy driving a small flock of sheep through the city in the early-morning light. There is great action to the piece. The photograph is the great recorder of life and all of life is here on this quiet Dublin street. There is no suffering in this image; it is totally different to *Anguish*. There is a joy or innocence to it, a feeling that makes me happy. Someone looking at this picture now might say that this was Dublin in a simpler time, but all times are tough. No, it is an image that shows

Shepherd and Sheep

the powerful connection that has now been lost between the urban and rural in Ireland. Here, in the 1960s, the two places are interlinked.

Ireland was not a quainter place back then – it was, in fact, a bigger place for many people. A journey to Dublin was a great event. It reminds me of the great drover journeys that were made from the countryside to the city back in that time. The stuff of the epic poems of Padraic Colum. When a sheep or cow being walked through Dublin was not an alien thing but a familiar sight. The days of the great cattle and sheep drives are changed now: in their place is the livestock truck; but those special trips were recorded in our art and remind us that Dublin once knew the farm as well as we rural people do. The drover men are not all gone either. Those who remain can still recall the fair days and the walk to Dublin and I have heard old men at the mart talk about these days from their boyhood.

Hartmann had fought in the Second World War and travelled the world before he came to Dublin to make this series of three thousand images. I like him because he was not a born photographer: he lived a full life before he became an artist; he worked in factories, soldiered and then eventually took up the camera. I think, in a way, he knew about humanity and life before he took his images and that is why they are so powerful.

You have to live a life before you can make art, at least that's my understanding. In ways, farming is like this too: you have to have some life experience to ground you in the work that you must do. Life, death, we deal with it all on the land, and having lived and seen how the world works can give us the tools to know how to farm better. Many young Irish farmers go to Australia or New Zealand these days and work on farms there and come back full of new ideas. It is part of the new rituals of this generation.

Hartmann said that a large portion of his work was concerned with people because people are the most inventive and news-making part of our lives. I think that, in this picture, he has captured something about our animal brothers too.

Each morning as I rise to go to the farm, I look at Hartmann's sheep and see all the potential before them and me. The picture says new things each day to me.

As I take another lamb in my hands to spray my mark upon his side, I think of Hartmann and those sheep. As I number the lamb and put a circle on his flank, I let him back to his mother. It is a painless job but an important one. He's got our marking now; he's got my hand upon him. It's my little moment of creativity, as small as it is.

Looking back now, I am glad I chose art all those years ago. It has enriched me and taught me so much. In art we can understand the world. In that comprehension I see the image and its lasting effect upon us. An image has the power to continue on in the mind for a lifetime. That is a powerful result for a silent, unmoving thing. I do not know what happened next to those sheep in Dublin, but I can always wonder and in that there is a great feeling of happiness.

I might, I think, put my Hartmann image in the shed, so it can remind me that these sheep are destined to go places far from here. That there is beauty in the quiet moments. Art practices – the writer, the poet, the filmmaker – are akin to being a shepherd because we share so many peaks and troughs; there are moments of joy and moments of sadness in projects. In farming, we can be a midwife and a funeral director sometimes all in one day, but it all has a meaning, as Pauline Matarasso says. There's a calling in that.

Health is wealth

The shed is full of lambs now. Everywhere I look, there is life around me. In the creep the older lambs play with their comrades and in the pens the newer arrivals prance and jump on their mothers' backs. One of the lambs in the creep has orf. It is a virus that produces scabs on their mouths. There is no cure for the disease: it comes and it goes as it has done most lambing seasons. We have disinfected the shed, sprayed our salves and chemicals, cleaned out pens but it is here now. There is a local woman who has the cure for the orf. One must call her and hold up the phone and she will say a prayer and the virus will pass from the animal. It is a form of magical thinking because we know the virus will fade in time by itself but we carry out the practice still. Her holy words are our best defence, and at times many farmers last ditch treatment. I have never spoken to her. I do not know what her prayers sound like.

Sometimes the orf is bad in the lambs and it prevents them from sucking from their mothers. Sometimes, too, the orf gets on the ewe's tits. This is dangerous, for the lamb could die of starvation and the ewe lose her tit to mastitis from not being milked. At times like this I have bottle-fed badly infected lambs to ensure they will get their feed into them. For the older lambs, cow's milk from a carton is fine and we heat it up in the microwave that sits in the shed. The lambs quickly drink the milk down and I know that they are at least not hungry. For the ewes, we must continue to milk out the infected tit, and in keeping the milk flowing we can ensure that the mastitis is kept at bay.

Magical thinking and medicine can combine on a farm, I suppose. We pray to gods and saints, we rub ointments, we call holy people. Lambing can be a season of magical thinking.

Health is something that I give great store to. The sick lamb has gotten me thinking about the body, my own and his.

When I came home all those years ago, I did not have a 'language of the body', as the Japanese writer Yukio

Mishima called it. I thought then that the mind was more important than the body. That the mind was the ultimate language. However, I see now that the body is integral. That the body, in short, is the bridge we must cross to come into communion with ourselves.

Mishima is a writer I have admired for many years, and while he has fallen out of fashion in the decades since his suicide by *seppuku* (the Japanese word for the ritual of cutting the belly with a sword) after a failed coup, I find his writing on the body to still be relevant. He referred to the body as the 'orchard' and believed that it was our job 'to cultivate the orchard'. It is why I have taken to gyms and pools and raceways and roads in the years since I returned to Ireland. In working my body, I have come to understand the language of the body, and I see now that the mind does not stand alone, that it needs arms and legs and hands and fingers. Even now, with all our tools and modern machinery, farming needs strength. Farming has taught me to build my body. A ram, a ewe, a cow and a horse need a robust-bodied farmer to handle them. Strength is needed on the farm to manoeuvre beasts and so I have worked to gain that strength. In turn, through the doing, it has allowed me to lift newborn calves over gates to clear fluid from their lungs or to hold a mare whose mind is concerned with bolting. In sun and steel I have worked the body, my body. And, as Mishima says, I have cultivated it to its capacity. I work now to never let the weeds take root in that orchard. There is a deep reason to that.

I attend the gym and pool six times a week, working on weights, on swimming and on cardio. It is my place to make sense of the day that has just been. For an hour or so, I push my body. There, in the gymnasium, is a place removed from the house and the farm, a place that offers me the joyful salve of sweat.

To a non-gym-goer, working the body this way might seem a strange activity. One goes to a gymnasium to lift iron in different patterns and cycles, grunting and repeating the tasks until a point of pain has been reached, or rather a point of understanding. The goer then carries out the same task again with another strange piece of equipment that focuses on some other isolated muscle. The tasks are carried out again and again throughout a week, and it might seem that with all the effort, we get nowhere. We do not arrive at any great destination but we do feel a sense of completion.

In the pool, which I attend most days too, I feel the sense of revelation. We uncover ourselves there as nowhere else, unfurling our near-naked bodies to our friends and strangers. It is a space where this is customary and accepted; it is part of the garb of the act. In the revelation we show our vulnerable, true selves.

The first gyms go back as far as ancient Greece and Persia, where physical education was deemed as important as mental education. The mind and the body worked in unison. Indeed, in Plato's *Republic*, he cites physical education and sport as a way to build virtue. The ideal citizen of classical life trained both body and mind. Plato goes as far as to say that it relates wholly to the health of the soul. The goal of exercise, for the puritan at least, should not just be the cultivation of physical beauty or the preparedness for the call to physical action (be it as a fighter or a solider), no, it should be the balance of the psyche. Namely the psyche of the soul and mind in the classical sense. The hardest victory, I suppose, is the victory over yourself.

I think about health a lot because seven years ago I did not have it. I had a mind that was suffering with a great depression and a body that was not in contact with its greater whole. Depression was my teacher, depression gave me an opportunity to change a life that was not working, and I came to see that it was, as Henri Nouwen said, 'a path to be a wounded prophet'. We can, in woundedness, live deeply and feel deeply. From a wounded place we can reach the wounds within others.

My seven years at home were a search through the door of darkness; they also were a door to spiritual education. It was during this time that prayer came into my life again. I rejected it at first, I pushed it aside, for what could prayer offer me when I had lost myself? Then I searched the world for answers. I looked to the east for tools. I looked to the west for guidance. Finally, I looked inside and found me.

My body, my psyche, united, I was reborn. What led to that rebirth? In ways it was the lambs. The lambs of that first season seven years ago put me back on the road to me. After I returned to the farm, they were there, coming into the world and needing my help, my hands to carry them safely across the threshold of life. They were a real and tangible thing that needed the care and attention that I could bring. To kiss a fluffy lamb's head when no one is watching is a joy. To untangle a ball of legs in the passage of a ewe and save a life is a thrill that not all can know. In the lambing season, we can be the shepherds we were always meant to be. It was the lambs who taught me that I could be a shepherd to my own life, that I could mend what had been broken.

I think, for all of us, there are episodes in life where we are called to be our higher selves. Where we are asked to answer a call. I tried in so many ways to reach that higher self: from asceticism to gluttony, from meditation to wild living. But then that is why we are here. To find out, to explore, to delve deep into the ocean of life with no thought of surfacing for air any time soon. A faith healer told me once that, after all those years of searching, I was meant to do a great thing in the world. He had seen it in a vision, he told me. I think now that great thing was these lambs of my own. To bring them into this world and to tell their story with the power of my words. The bell of life rang out through the darkness, my darkness.

Maybe being a shepherd is my true calling on earth.

Father Sean told me some time ago about a close friend of his who came to understand the power of life, or the story of life, through illness. This man had suffered from various illnesses throughout his life and was frail even in his best young days. He, too, was a journeyman through life who came to see the beauty in broken things. He travelled to the far corners of the world like I did and found himself, I think. I think of that man often and how the land, how the earth, is a prayer.

It's in sickness that I think we can come to truly understand the value of health and, in doing that, we can understand the soul and life. I take joy in my strength on the farm. I take solace in being well, for it was not always so. Being able to handle the rams or bull calves makes me feel alive, makes me feel a full-spirited being. It is part of the prayer of the body, the prayer I woke up to seven years ago.

The lamb with the orf cries now and I put aside my thoughts and pick him up and inspect him. Part of me would love to pick away the scabs that have formed around his mouth but that will not end the disease. We share a look now in this intimate moment. It is not a look of superiority, for I need him as much as he now needs me in his illness. I read his number and find his mother and put them in isolation in a pen of their own in the hope that he won't give orf to the other lambs. I care for the health of these creatures, my creatures. To care for another creature on this planet requires, I think, a great deal of love. That is the basic root of being a shepherd, as I have learned over this intimate season.

A healthy, fresh life for the lambs and their mothers is much better than all the industrial medicines in the world. So often, in farms around the world, we farmers do not know what these medicines truly do. We administer them and an animal lives or dies. There is seldom an in-between. The health of these sheep, these lambs, is important to me. I do not wish to fill them full of drugs or chemicals. I will resort to medicine when it is truly needed, of course, but we must be careful in our use of these same medicines. Already, over half of all antibiotics given in the world are administered not to people but to animals, leading to antibiotic resistance. Drugs are overused in animal husbandry in some parts of the world. There have been calls from on high to curb the use of these drugs for fear that it will fuel what has been called a silent pandemic. A statement from the G7 summit in 2021 said that in some parts of the world antibiotics are given to animals who are not sick as a preventative and to promote growth. As many media outlets reported following this statement, so much of this overuse of drugs comes from intensive agriculture. Thankfully, to date, Ireland is amongst the lowest users of antibiotics in livestock in the EU, according to the *Irish Farmers Journal* in 2017.

In his climate encyclical, Pope Francis wrote that the fate of creation is tied to the fate of all humanity. Even the spiritual leader has talked about the dangers to health of pesticides and other irresponsible chemicals leading to illnesses in children. He says that we are mortgaging our future while getting ever sicker because we are not in unity with our home. For millions, the decision to eat meat, to eat these lambs sometime in the future, is entirely divorced from the journey I have been on with them. People do not know the process or road it has been on, nor in the cases of some crops what chemicals have been sprayed upon them. As the environmentalist Rachel Carson pointed out many years ago, we are in contact with dangerous chemicals in our world now from birth until death.

Animals must die so that we can eat. For those who eat meat it can be done in a sustainable way, but we must be careful not to abuse our relationship with animals and the earth in the doing of this. Cheap meat coming from, say, deforested parts of Brazil is having a huge impact on the earth. We would do well to remember that. It does not have to be all-or-nothing but there must be balance.

For an animal to be healthy, in this case the lamb I am holding, they need to be with their mothers, they need fresh grass, and they need the milk of life. That is how *their* psyches can be in order. When we share our look across the animal divide, I can tell from his signals what he needs. It is a look that has been learned between humanity and sheep since this dance began all those thousands of years ago. My soul, his animal soul, they've met before; they know the paths they are going to take.

All these signs tell me one big thing: that the final step is coming, that we must soon let the animals out of the shed. The good weather is here. The fields full of grass are calling for them and, at last, the lambing season, the journey, is coming to an end. Health, true health, is the wealth that I wish to give to these lambs. To do that, I must let them be what they were always meant to be: sheep.

A sick lamb has taught me that. I am, at last, becoming the shepherd I hoped to be.

LESSON 12

Love is what you need

The day has arrived, and the time is right. We can wait no longer. I'm in the sheep shed, looking at the mothers and babes. They are fit and strong and ready to take their next steps in the wider world. The shed has shielded them from the weather of the world outside, but like a ship in harbour, they are destined for bigger things, for greater things. The fields, the poems of our lives, are calling and in the fresh grass there is a new destination. The seasons have changed; winter has given way to spring. The sheep are that prism through which I can see my future. It has always been this way. It will always be this way. There is a new feeling in the air.

For me, it has been a time of a quiet personal evolution, from man to shepherd. From writer to true farmer. When I was young, I dreamed of words; now, I dream of sheep. I chose this dream. All this long lambing season I have been thinking of a play I read once, *Kicking a*

Dead Horse by the American playwright Sam Shepard. The work is about a man attempting to bury a horse that has died on him, and he is stranded in the desert of the old west. I had thought of the play before, when I was building the square hay bales in the shed last summer. I thought of it because it seemed an endless job and the hay bales kept falling on me. Sometimes on the farm we seem to perform mindless work that goes on forever or into eternity, whichever is the longer. Shepard's play, while about the burying of the horse, is also about the work of continuing. Of never giving up. That in remaining constant to our calling, we can become the people we were always destined to be.

In picking up the collapsing bales again and again, I thought of the futility of the task. I was not a builder, I reasoned, or rather, I realized, but like Shepard's protagonist, I kept going. I kept building, even though I knew it was beyond me. In our work on the farm, we must keep going, and we continue. We work the land, we work the soil, we work the animals. We have the never-giving-up spirit in our blood, which has passed down through our ancestors to this moment. When I continue, I keep the memory of all my people alive; they are the unquenched candle.

Walking the shed now, picking up the sweet lambs in my hands, I see that continuing has made me a better man. That the sheep have been teachers, that the lambs have been the reward for remaining steadfast. It has been a months-long lesson that has changed me forever.

Shepard's play is also about the search for authenticity. The central character, Hobart Struther, is a man who made his money buying art cheap and selling it for exorbitant prices in New York. But he feels himself lost and has come out west to find himself again. In his search for authenticity, he has come home to understand the landscape that made him. I can see now that in coming home seven years ago I, too, have been on a journey of authenticity. It has not been a simple journey, but then no great journeys of the heart are.

I have travelled so far in these months, and yet I have not left the farmyard. I have been a place traveller, a time traveller, a song traveller. I have remembered all my lives; maybe it was the sheep that made this so. The space that I have occupied has shown me that sometimes the story is the destination. The journey is what makes us richer at its end.

I put down the lamb and walk towards the shed door. It has not been opened in a long time. It is the door to the future. I see now that, unlike Struther, I have something in my heart that was wholly absent in him. I have that simple but overlooked thing: love. I carried out my building of the hay that day last summer because I wanted to have food in place for the animals when the winter came. That same hay I have so carefully fed to my sheep to see them through their days with us on this farm. That hay has given sustenance to them all. It was love and affection that drove me on so that day could come to this one.

I have said that the sheep, the lambs, were a stake in the future here at Birchview, but they are also a stake in the future of my heart. It's only at the end of this journey that I see that love has been carrying me. Through the late nights, through the bad luck, through the lessons. Love, I think, is something that drives us onwards toward the future. Life acquires meaning through love, as Hermann Hesse once wrote. 'The more we are capable of love and dedication, the more our lives will be rich with meaning.'

Maybe love is meaning. All nature has a feeling; perhaps that feeling is love. I think I am coming to know its presence, its warmth.

Some would ask if we can love an animal. But my answer now is yes: they are a part of our family, part of my family. I have grown in the company of these sheep, and though the lambs will one day go to the slaughterhouse to pay for these labours, I have a love for them in my heart, too. They will earn me a living, and there is an honesty in that. An age-old act of the earth.

When we work with or in love, great things can be achieved. We can surmount the layers of life and death. In our love for all the creator's creatures, we find ourselves.

Before I open the door, I make a final check of the flock, my flock. I inspect the mothers and the lambs. The sickness is gone from them, the spray-painted numbers are visible and intact. I count them all and count them again to be sure. They are all here, all present, all ready. I have bound myself to these sheep, and now that the season is over, all that is left to do is one final thing: release them. It has been a lambing season I will not forget because it has been my own.

I open the shed door and walk up to the yard gate, over the bridge where the river runs deep and the fields lie in their promise. I make my calls now, and the mothers that my flock have become know what this means: that the lambing is over and that they must make their way to the fields and take care of themselves.

The ewes and their lambs come running now and breathe in the fresh air. In their baas and calls, I feel a sense of excitement. They were born to this life, and in the jumps of the lambs in the air, I smile. Not all life's lessons are like this. Not all are so sweet. As they run by me, I feel a sense of pride but also, in a way, a sense of being at one with things. I close the yard gate and walk behind them as we make our way to the reseeded field.

There is hope in my steps now, for hope is the great companion to love. In hope, we can be our better selves in this world. In hope, I put my future. Hope that the lambs will thrive, that the sheep will be good mothers, hope that I shall get a fair price. And hope that life will continue.

The making of words, like the making of twelve sheep, does not happen overnight. I have given my labours to the earth and to the page. I see now through this work that I have emerged out the other side as a changed person. We have come through the season, they and I. We must all of us pick up the fallen bales of life, again and again, making a haystack of living that is tight and strong. Love and time have brought us this far. Now it is up to nature.

That is my last lesson.

Epilogue

The season has ended, though I have not and nor have the sheep. We push on now through the good spring days, through to where we can earn a living from these sheep. The twelve have taught me. In ways, I am a wiser man, and my decision to stay on the land has been the right one. Walking through the lush grass now, I know that my soul tiredness is gone. It has been replaced with a weariness brought about by hard labour, and yet that has made me bloom. It is a pleasing tiredness.

Lessons, like souls, must be discovered and earned. We can all of us find our teachers in life. Perhaps they are right in front of us, as they have been for me. I am without words in this moment, but some things go beyond words. Sometimes words are not needed. We are not free until we live in the joyful moment. I have earned my place on the land at long last. I have become the farmer and perhaps the writer I always sought to be. It took twelve sheep to make that possible. It took a lambing season to bring me back to me.

I have come to know the Pachamama, or Mother Earth, through this journey with these animals. Each of the sheep has taught me something about myself. Most of all, they have taught me how to be alive again.

There are lessons here for all of us, I think. We have but to listen to nature and she will guide us on how to be a shepherd to ourselves in our days upon this earth. There is grace for all who look. There is grace for all who listen.

ABOUT THE AUTHOR

John Connell is a multi award-winning author, journalist, playwright, documentary producer, motivational speaker and farmer. His documentary programs have won over a dozen international awards and been screened around the world, and he is a regular contributor to BBC Radio 4's *A Point of View*. His number-one bestselling memoir *The Cow Book* was awarded Popular Non Fiction Book of the Year at the Irish Book Awards. He is the co-founder and co-director of the Granard Booktown Festival, which takes place in Ireland's national booktown. He lives and farms in County Longford in the Irish midlands where he works on his writing and film projects.